5 STEPS TO A 5

500

AP Calculus AB/BC Questions
to know by test day

WITHDRAWN

Zachary Miner
Lena Folwaczny

Mc Graw Hill

New York Chicago San Francisco Lisbon London Madrid Mexico City
Milan New Delhi San Juan Seoul Singapore Sydney Toronto

The McGraw·Hill Companies

ZACHARY MINER holds a Ph.D. in mathematics from the University Texas at Austin. He taught calculus at the undergraduate level for seven years.

LENA FOLWACZNY holds an M.A. in pure mathematics from the University of Illinois in Chicago. She is an experienced AP Calculus and math tutor, and she has taught calculus at the undergraduate level since 2006.

Copyright © 2012 by The McGraw-Hill Companies, Inc. All rights reserved. Printed in the United States of America. Except as permitted under the United States Copyright Act of 1976, no part of this publication may be reproduced or distributed in any form or by any means, or stored in a database or retrieval system, without the prior written permission of the publisher.

1 2 3 4 5 6 7 8 9 10 11 12 13 14 15   QFR/QFR   1 9 8 7 6 5 4 3 2 1

ISBN   978-0-07-175370-8
MHID   0-07-175370-2

e-ISBN   978-0-07-175371-5
e-MHID   0-07-175371-0

Library of Congress Control Number 2010935988

Trademarks: McGraw-Hill, the McGraw-Hill Publishing logo, 5 Steps to a 5, and related trade dress are trademarks or registered trademarks of The McGraw-Hill Companies and/or its affiliates in the United States and other countries and may not be used without written permission. All other trademarks are the property of their respective owners. The McGraw-Hill Companies is not associated with any product or vendor mentioned in this book.

AP, Advanced Placement Program, and *College Board* are registered trademarks of the College Entrance Examination Board, which was not involved in the production of, and does not endorse, this product.

Series interior design by Jane Tenenbaum.

McGraw-Hill books are available at special quantity discounts to use as premiums and sales promotions or for use in corporate training programs. To contact a representative, please e-mail us at bulksales@mcgraw-hill.com.

This book is printed on acid-free paper.

CONTENTS

INTRODUCTION

Congratulations! You've taken a big step toward AP success by purchasing *5 Steps to a 5: 500 AP Calculus AB/BC Questions to Know by Test Day*. We are here to help you take the next step and score high on your AP Exam so you can earn college credits and get into the college or university of your choice.

Thus book gives you 500 AP-style questions—both multiple-choice and free-response—that cover all the most essential course material. Each question has a detailed answer explanation. These questions will give you valuable independent practice to supplement your regular textbook and the groundwork you are already doing in your AP classroom.

This and the other books in this series were written by expert AP teachers who know your examination inside out and can identify the crucial examination information as well as questions that are most likely to appear on the examination.

You might be the kind of student who takes several AP courses and needs to study extra questions a few weeks before the examination for a final review. Or you might be the kind of student who puts off preparing until the last weeks before the examination. No matter what your preparation style is, you will surely benefit from reviewing these 500 questions, which closely parallel the content, format, and degree of difficulty of the questions on the actual AP examination. These questions and their answer explanations are the ideal last-minute study tool for those final few weeks before the test.

Remember the old saying, "Practice makes perfect." If you practice with all the questions and answers in this book, we are certain you will build the skills and confidence needed to do great on the exam. Good luck!

—The Editors of McGraw-Hill Education

NOTE: The questions designated **BC** in the following pages cover topics tested only on the Calculus BC examination.

CHAPTER 1

Limits and Continuity

1. What is $\lim_{x \to \frac{\pi}{4}} \tan x$?

   (A) -1

   (B) 0

   (C) 1

   (D) $\dfrac{\sqrt{2}}{2}$

   (E) $\dfrac{\sqrt{3}}{2}$

2. Find the limit: $\lim_{t \to -3} \dfrac{t+3}{t^2+9}$.

   (A) $-\dfrac{1}{3}$

   (B) 0

   (C) $\dfrac{1}{3}$

   (D) 1

   (E) The limit does not exist.

3. Find the limit: $\lim_{x \to 3} \dfrac{x^3 + 2x^2 - 9x - 18}{x - 3}$.

   (A) 6

   (B) 30

   (C) 1

   (D) ∞

   (E) The limit does not exist.

4. Find the limit: $\lim_{t \to 0} \dfrac{\sqrt{4-t} - 2}{t}$.

   (A) ∞

   (B) $\dfrac{1}{4}$

   (C) $-\infty$

   (D) $-\dfrac{1}{4}$

   (E) The limit does not exist.

5. Find the limit: $\lim_{s \to 9} \dfrac{9 - s}{\sqrt{s} - 3}$.

   (A) ∞

   (B) $-\infty$

   (C) -6

   (D) 0

   (E) The limit does not exist.

6. Find the limit: $\lim_{x \to \infty} \dfrac{x + 4}{x^2 + 16}$.

   (A) ∞

   (B) $\dfrac{1}{4}$

   (C) 1

   (D) 0

   (E) The limit does not exist.

7. Find the limit: $\lim_{\theta \to -\infty} \dfrac{\sin \theta}{\theta}$.

   (A) 0

   (B) 1

   (C) -1

   (D) $-\infty$

   (E) The limit does not exist.

8. Find the limit: $\lim_{t \to \infty} \dfrac{\sqrt[3]{t^3 - 8}}{2t}$.

   (A) 0

   (B) $\dfrac{1}{2}$

   (C) 1

   (D) 4

   (E) The limit does not exist.

9. Find the limit: $\lim_{x \to 0} \dfrac{4x^2}{1 - \cos 2x}$.

(A) 0

(B) 2

(C) 4

(D) 16

(E) The limit does not exist.

10. Find the limit: $\lim_{\theta \to 0} \dfrac{\cos^2 \theta - 1}{\theta \cos \theta + \theta}$.

(A) −1

(B) 0

(C) $\dfrac{\sqrt{2}}{2}$

(D) 1

(E) The limit does not exist.

11. Find the limit: $\lim_{y \to 0} \dfrac{(\tan y)(\cos y)}{y}$.

(A) −1

(B) 0

(C) $\dfrac{\sqrt{2}}{2}$

(D) 1

(E) The limit does not exist.

12. Find the limit: $\lim_{z \to -1} \dfrac{z^3 + 1}{z + 1}$.

(A) −1

(B) 0

(C) 1

(D) 3

(E) The limit does not exist.

13. Find the horizontal asymptote(s) of $f(t) = \dfrac{27t - 18}{3t + 8}$.

(A) $y = 9$

(B) $y = 6$

(C) $y = -\dfrac{9}{4}$

(D) $y = -6$

(E) There are no horizontal asymptotes.

14. Find the vertical asymptote(s) of $f(x) = \dfrac{x^2 + 2x + 1}{x^2 - 1}$.

   (A) $x = 1$
   (B) $x = -1$
   (C) $x = 1$ and $x = -1$
   (D) $y = 1$
   (E) $y = -1$

15. For what value of h is $f(x) = \begin{cases} \dfrac{6x^2 - 11x - 10}{2x - 5}, & x \neq \dfrac{5}{2} \\ h, & x = \dfrac{5}{2} \end{cases}$ continuous at $x = \dfrac{5}{2}$?

   (A) -3
   (B) 0
   (C) 3
   (D) $\dfrac{25}{2}$
   (E) $\dfrac{19}{2}$

16. For what value of k is $g(x) = \begin{cases} 2x + 7, & x \leq 4 \\ -3x - k, & x > 4 \end{cases}$ a continuous function?

   (A) 0
   (B) 4
   (C) -4
   (D) 27
   (E) -27

17. Find the value of m for which $h(x) = \begin{cases} 5x - 13, & x \leq 2 \\ x^2 - 7x + m, & x > 2 \end{cases}$ is a continuous function.

   (A) $\dfrac{7 - \sqrt{37}}{2}$
   (B) -3
   (C) 3
   (D) 7
   (E) $\dfrac{7 + \sqrt{37}}{2}$

18. Find the point(s) of discontinuity of the function $f(x) = \dfrac{3x + 1}{2x^3 - 8x^2 - 64x}$.

   I.   0
   II. 4
   III. 8

   (A) I only
   (B) II only
   (C) III only
   (D) I and II only
   (E) I and III only

19. On which interval(s) is the function $g(x) = \begin{cases} \dfrac{x^2 + 4x - 21}{x^2 - 8x + 15}, & x \neq 3, 5 \\ -5, & x = 3 \\ -\dfrac{7}{5}, & x = 5 \end{cases}$
is continuous?

   I. $(-\infty, 3)$
   II. $(3, \infty)$
   III. $(5, \infty)$

   (A) I only
   (B) II only
   (C) III only
   (D) I and II only
   (E) I and III only

20. Let $h(x)$ be continuous on $[-2, 3]$ with some of the values shown in the following table:

x	-2	0	3
$h(x)$	7	a	5

   If $h(x) = 4$ has no solutions on the interval $[-2, 3]$, which of the following values are possible for a?

   I. -1
   II. 4
   III. 6

   (A) I only
   (B) II only
   (C) III only
   (D) I and II only
   (E) I and III only

21. For the function $f(x) = \dfrac{x^3 - 6x^2 + 11x - 6}{x^3 - 7x + 6}$, which point of discontinuity is not removable?

   (A) $x = -3$
   (B) $x = 1$
   (C) $x = 2$
   (D) $x = 3$
   (E) $x = 6$

22. What is $\lim_{\theta \to \frac{\pi}{4}} \dfrac{\cos 2\theta}{\cos \theta - \sin \theta}$?

(A) 0

(B) $\dfrac{\sqrt{2}}{2}$

(C) 1

(D) $\sqrt{2}$

(E) The limit does not exist.

23. Which represents the removable discontinuity of the function $f(t) = \dfrac{t^4 - 2t^3 - 13t^2 + 14t + 24}{t^4 - 2t^3 - 13t^2 + 38t - 24}$?

(A) 1

(B) 2

(C) 3

(D) 4

(E) 6

24. Find the vertical asymptote of $f(x) = \dfrac{x^2 + x - 12}{x^2 - 7x + 12}$.

(A) $x = -4$

(B) $x = -3$

(C) $x = 0$

(D) $x = 3$

(E) $x = 4$

25. For what value of k is the function $f(x) = \begin{cases} \dfrac{6x^2 + 5x - 56}{2x + 7}, & x \neq -\dfrac{7}{2} \\ k, & x = -\dfrac{7}{2} \end{cases}$ continuous?

(A) $-\dfrac{37}{2}$

(B) $-\dfrac{7}{2}$

(C) $\dfrac{5}{2}$

(D) 3

(E) $\dfrac{37}{2}$

26. For what value of h is the function $g(t) = \begin{cases} 2x^2 - 8x + 7, \, t \geq 5 \\ 2x^3 - 9x^2 - 2x + h, \, t < 5 \end{cases}$ continuous?

   (A) −17
   (B) −5
   (C) 0
   (D) 2
   (E) 32

27. For the function $(x) = \dfrac{x^3 + 5x^2 - 2x - 24}{x^3 - 3x^2 - 10x + 24}$, which point of discontinuity is not removable?

   (A) 0
   (B) 1
   (C) 2
   (D) 3
   (E) 4

28. On which interval(s) is the function $g(t) = \begin{cases} \dfrac{4t^2 - 8t - 21}{4t^2 - 20t + 21}, \, t \neq \dfrac{3}{2}, \dfrac{7}{2} \\ -1, \, t = \dfrac{3}{2} \\ \dfrac{5}{2}, \, t = \dfrac{7}{2} \end{cases}$ continuous?

   I. $\left(-\infty, \dfrac{3}{2}\right)$

   II. $\left(-\infty, \dfrac{7}{2}\right)$

   III. $\left(\dfrac{3}{2}, \infty\right)$

   (A) I only
   (B) II only
   (C) III only
   (D) I and II only
   (E) I and III only

29. Find the horizontal asymptote(s) of $f(x) = \dfrac{x}{x^2 + 1}$.

   I. $x = -1$
   II. $x = 0$
   III. $x = 1$

   (A) I only
   (B) II only
   (C) III only
   (D) I and II only
   (E) I and III only

30. Which represents the removable discontinuity of the function
$(t) = \dfrac{x^4 - 7x^3 + 5x^2 + 31x - 30}{x^4 + x^3 - 19x^2 + 11x + 30}$?

    (A) 1
    (B) 2
    (C) 3
    (D) 5
    (E) 7

31. For what value of a is the function $g(x) = \begin{cases} \dfrac{x^4 - x^3 + x - 1}{x - 1}, & x \neq 1 \\ a, & x = 1 \end{cases}$ continuous?

    (A) 0
    (B) 1
    (C) 2
    (D) 6
    (E) 7

32. For the function $(x) = \dfrac{2x^3 + x^2 - 25x + 12}{2x^3 + 3x^2 - 23x - 12}$, which point of discontinuity is not removable?

    (A) $x = -4$
    (B) $x = -3$
    (C) $x = -\dfrac{1}{2}$
    (D) $x = 3$
    (E) $x = 4$

33. Find the value of b so that $f(x) = \begin{cases} \dfrac{\sqrt{x^2 + 1} - 1}{x}, & x \neq 0 \\ b, & x = 0 \end{cases}$ is continuous.

    (A) $-\infty$
    (B) -1
    (C) 0
    (D) 1
    (E) ∞

34. Find the value of c so that $h(t) = \begin{cases} \dfrac{t^2 - 4}{t^3 - 8}, & t \neq 2 \\ c, & t = 2 \end{cases}$ is continuous.

    (A) 0
    (B) $\dfrac{1}{3}$
    (C) $\dfrac{1}{2}$
    (D) 1
    (E) The discontinuity at $t = 2$ is not removable.

35. Find the value of k so that $g(x) = \begin{cases} \dfrac{x^2+1}{x+1}, & x \neq -1 \\ k, & x = -1 \end{cases}$ is continuous.

(A) −1
(B) 0
(C) 1
(D) 2
(E) The discontinuity at $x = -1$ is not removable.

36. Let $f(x)$ be continuous on $[-5, -2]$ with some of the values shown in the following table:

x	−5	−4	−2
$h(x)$	−11	a	−3

If $f(x) = -1$ has no solutions on the interval $[-5, -2]$, which of the following values are possible for a?

I. −4
II. −3
III. 0

(A) I only
(B) II only
(C) III only
(D) I and II only
(E) I and III only

37. On which interval(s) is the function $g(x) = \begin{cases} \dfrac{x^2 - 2x - 24}{x^2 + 10x + 24}, & x \neq -4, -6 \\ -5, & x = -4 \\ -1, & x = -6 \end{cases}$ continuous?

I. $(-\infty, -4)$
II. $(-4, 6)$
III. $(-4, \infty)$

(A) I only
(B) II only
(C) III only
(D) I and II only
(E) I and III only

38. Find the limit: $\lim_{x \to 0} \dfrac{\sqrt{6-x} - \sqrt{6}}{x}$.

(A) $-\infty$

(B) $-\dfrac{\sqrt{6}}{12}$

(C) 0

(D) $\dfrac{1}{2\sqrt{6}}$

(E) ∞

39. Find the limit: $\lim_{x \to 11} \dfrac{\sqrt{x+5} - 4}{x - 11}$.

(A) $-\infty$

(B) $-\dfrac{1}{8}$

(C) 0

(D) $\dfrac{1}{8}$

(E) ∞

40. Find the limit: $\lim_{x \to 0} \dfrac{\sin x^3}{x^2}$.

(A) $-\infty$

(B) 0

(C) $\dfrac{\sqrt{2}}{2}$

(D) 1

(E) ∞

41. Find the limit: $\lim_{x \to 3} \dfrac{x^2 - x}{x - 3}$.

(A) $-\infty$

(B) -1

(C) 0

(D) 1

(E) ∞

42. Find $\lim_{x \to \infty} \dfrac{45x^2 + 13x - 18}{4x - 9x^3}$.

(A) $-\infty$

(B) -5

(C) 0

(D) 5

(E) ∞

43. For the function $f(x) = \begin{cases} \dfrac{28x^2 - 13x - 6}{7x + 2}, & x \neq -\dfrac{7}{2} \\ k, & x = -\dfrac{7}{2} \end{cases}$ what must be the

value of k for the function to be continuous at $x = -\dfrac{7}{2}$?

(A) $-\dfrac{23}{3}$

(B) $-\dfrac{7}{2}$

(C) 0

(D) $\dfrac{23}{3}$

(E) 4

44. Find $\lim_{x \to 5} \dfrac{x^2 - 24}{5 - x}$.

(A) $-\dfrac{24}{5}$

(B) -1

(C) 0

(D) 1

(E) The limit does not exist.

45. Find $\lim_{x \to \infty} \dfrac{\sqrt{x} - 7}{6 - 5\sqrt{x}}$.

(A) $-\infty$

(B) $-\dfrac{7}{6}$

(C) $-\dfrac{1}{5}$

(D) 0

(E) ∞

46. (A) Find $\lim_{x \to \frac{\pi}{4}} \dfrac{x(1 - \tan x)}{\cos x - \sin x}$.

(B) Use the result of (a) to derive an approximation of $\sec x$ in terms of x for values of x near 0.

47. (A) Use a limit to verify the formula for the area of a circle using inscribed n-gons.

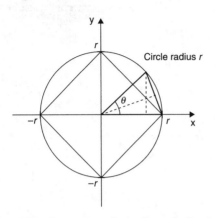

    (B) Use a limit to verify the formula for the area of a circle using circumscribed n-gons.

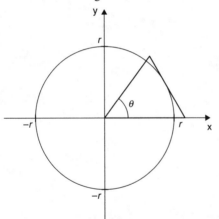

48. (A) Let $f(x) = \dfrac{p(x)}{q(x)}$, where p and q are both 4th degree polynomial functions. Discuss the possible number of continuities, removable and nonremovable.

    (B) How does the situation change if q is a 5th degree polynomial?

    (C) What is the situation regarding continuities if p is a 5th degree polynomial, but q is a 4th degree polynomial?

49. Many junior high students have noticed that $1 = \dfrac{1}{3} + \dfrac{2}{3} = 0.\overline{33} + 0.\overline{66} = 0.\overline{99}$. How would you use a limit to prove the same result?

50. Find the limit: $\displaystyle\lim_{x \to 0} \dfrac{4 - \sqrt{16 - x}}{x}$.

CHAPTER 2

Differentiation

51. What is $\lim_{\Delta x \to 0} \dfrac{\sin\left(\dfrac{\pi}{4} + \Delta x\right)\cos\left(\dfrac{\pi}{4} + \Delta x\right) - \sin\left(\dfrac{\pi}{4}\right)\cos\left(\dfrac{\pi}{4}\right)}{\Delta x}$?

(A) -1

(B) 0

(C) $\dfrac{1}{2}$

(D) 1

(E) $\dfrac{\sqrt{3}}{2}$

52. What is $\lim_{h \to 0} \dfrac{\tan\left(\dfrac{\pi}{4} + h\right) - \tan\left(\dfrac{\pi}{4}\right)}{h}$?

(A) 0

(B) $\dfrac{\sqrt{2}}{2}$

(C) 1

(D) $\dfrac{\sqrt{3}}{2}$

(E) 2

53. If $f(x) = \sin(x) \cdot \cos^2(\pi - x)$, then $f'\left(\dfrac{\pi}{4}\right) =$

(A) $-\dfrac{\sqrt{2}}{4}$

(B) 0

(C) $\dfrac{\sqrt{2}}{4}$

(D) $\dfrac{\sqrt{2}}{2}$

(E) $\dfrac{3\sqrt{2}}{4}$

54. What is the slope of the tangent to the curve $3x^2 + y^3 = -37$, when $x = 3$?

(A) $-\dfrac{8}{3}$

(B) -1

(C) $-\dfrac{3}{8}$

(D) 1

(E) $\dfrac{8}{3}$

55. What is the slope of the tangent to the curve $\cos(x) + \dfrac{y^2}{2} = 1$, when $x = 0$?

(A) -1

(B) 0

(C) $\dfrac{1}{2}$

(D) 1

(E) Undefined

56. Find $\dfrac{dy}{dx}$ if $\sec y = (y - x)^3$.

(A) $\dfrac{y - x}{3 \sec y}$

(B) $\dfrac{3x - 3y}{\tan y + 3x - 3y}$

(C) $\dfrac{3x - 3y}{\sec y + 3x - 3y}$

(D) $\dfrac{3x^2 - 6xy + 3y^2}{\sec y \cdot \tan y + 3x^2 - 6xy + 3y^2}$

(E) $\dfrac{3x - 3y}{\sec y \cdot \tan y + 3x - 3y}$

57. Find $\dfrac{dy}{dx}$ if $y = 7 + 5^{x^2 + 2x - 1}$.

(A) $5^{x^2 + 2x - 1}$

(B) $(5^{x^2 + 2x - 1})(2x + 2)$

(C) $(x^2 + 2x - 1)(5^{x^2 + 2x - 1})$

(D) $(2x^3 + 6x^2 + 2x - 2)(5^{x^2 + 2x - 1})$

(E) $(x^2 + 2x - 1)(2x + 2)$

58. The $\lim_{h \to 0} \dfrac{\ln(2x + h) - \ln(2x)}{h}$ is

(A) $\dfrac{1}{x}$

(B) $\dfrac{1}{2x}$

(C) $\ln x$

(D) $\ln(2x)$

(E) Undefined

59. The slope of the line normal to the graph of $r = 2 \cos\left(\dfrac{\theta}{2}\right)$ at $\theta = \pi$ is

(A) -1

(B) 0

(C) 1

(D) 2

(E) Undefined

60. If $f(5) = 6$ and $f'(5) = 7$, then the equation of the tangent to the curve $y = f(x)$ at $x = 5$ is

(A) $y = 7x - 35$
(B) $y = 7x - 29$
(C) $y = 6x - 42$
(D) $y = 6x - 37$
(E) $y = 6x - 7$

Table for Questions 61 and 62

x	1	2	3
f(x)	3	0	1
f'(x)	-3	5	-2
g(x)	4	-1	1
g'(x)	-4	3	0

61. If $h(x) = f(x)g(x)$, then $h'(2) =$

(A) -5
(B) -3
(C) 0
(D) 3
(E) 15

62. If $h(x) = g(f(x))$, then $h'(3) =$

(A) -3
(B) 0
(C) 8
(D) 9
(E) 12

63. If $2x^4 - xy + 3y^3 = 12$, then, in terms of x and y, $\dfrac{dy}{dx} =$

(A) $\dfrac{8x^3 - y}{y^2 - 9x}$

(B) $\dfrac{8x^3 - y}{9y^2 - x}$

(C) $\dfrac{y - 8x^3}{9y^2 - x}$

(D) $\dfrac{y - 8x^3}{x - 9y^2}$

(E) $\dfrac{8y - x^3}{x - 9y^2}$

64. If $f(x) = 4\sec^3(5x)$, then $f'(x)$ is

(A) $12\sec^2(5x)$
(B) $60\sec^2(5x)$
(C) $60\tan(5x)$
(D) $60\sec^3(5x)\tan(5x)$
(E) $60\sec^2(5x)\tan^2(5x)$

65. If $y = \dfrac{e^{3x^2}}{6}$, then $y''(0) =$

(A) $6x^2 \cdot e^{3x^2}$
(B) $e^{3x^2} + 6x^2 \cdot e^{3x^2}$
(C) $6x^2 + e^{3x^2}$
(D) $x \cdot e^{3x^2}$
(E) $x + e^{3x^2}$

66. If $y = \dfrac{2x+7}{5-2x}$, then $\dfrac{dy}{dx} =$

(A) $\dfrac{24}{(5-2x)^2}$

(B) $\dfrac{2x+7}{(5-2x)^2}$

(C) $\dfrac{8x+4}{(5-2x)^2}$

(D) $\dfrac{25-4x^2}{(5-2x)^2}$

(E) $\dfrac{4x^2-25}{(5-2x)^2}$

67. If $f(x) = \ln\left(\dfrac{1}{x}\right)$, then $f'(x) =$

(A) $-\dfrac{\ln\left(\dfrac{1}{x}\right)}{x^2}$

(B) $-\dfrac{1}{x}$

(C) $\dfrac{1}{x}$

(D) $\dfrac{\ln\left(\dfrac{1}{x}\right)}{x}$

(E) $\dfrac{\ln\left(\dfrac{1}{x}\right)}{x^2}$

68. If $f(x) = -\cos^2(x^2 + 2x - 3)$, then $f'(x) =$

(A) $-2\cos(x^2 + 2x - 3)$
(B) $-(4x + 4)\cos(x^2 + 2x - 3)$
(C) $(4x + 4)\cos(x^2 + 2x - 3)\sin(x^2 + 2x - 3)$
(D) $(4x + 4)\sin(x^2 + 2x - 3)$
(E) $2\sin(x^2 + 2x - 3)$

69. If $y = \dfrac{e^{x^2}}{x}$, then $y' =$

(A) $2e^{x^2}$

(B) $\dfrac{e^{x^2}}{x}$

(C) $\dfrac{e^{x^2}}{2x}$

(D) $\dfrac{e^{x^2}}{x^2}$

(E) $\dfrac{e^{x^2}(2x^2 - 1)}{x^2}$

70. What is the slope of the line that is tangent to $f(x) = \ln(\arcsin(x))$ at $x = \dfrac{\sqrt{2}}{2}$?

   (A) $\dfrac{\pi}{4}$

   (B) $\dfrac{\pi}{8}$

   (C) $\dfrac{4}{\pi}$

   (D) $\dfrac{8}{\pi}$

   (E) $\dfrac{4}{\pi^2}$

71. What is

$$\lim_{\Delta x \to 0} \frac{\sin\left(\dfrac{\pi}{4} + \Delta x\right)\cos\left(\dfrac{\pi}{4} + \Delta x\right)\ln\left(\dfrac{\pi}{4} + \Delta x\right) - \sin\left(\dfrac{\pi}{4}\right)\cos\left(\dfrac{\pi}{4}\right)\ln\left(\dfrac{\pi}{4}\right)}{\Delta x}?$$

   (A) -1

   (B) $-\dfrac{2}{\pi}$

   (C) $-\dfrac{1}{2}$

   (D) $\dfrac{2}{\pi}$

   (E) $\ln\left(\dfrac{\pi}{4}\right) + \dfrac{2}{\pi}$

72. What is $\lim_{h \to 0} \dfrac{\sec\left(\dfrac{4\pi}{3} + h\right) - \sec\left(\dfrac{4\pi}{3}\right)}{h}$?

   (A) $-2\sqrt{3}$

   (B) $-\dfrac{\sqrt{3}}{2}$

   (C) 0

   (D) $\dfrac{\sqrt{3}}{2}$

   (E) $2\sqrt{3}$

73. If $f(x) = \cos(3x) \cdot \sin^2(2x - \pi)$, then $f'\left(\dfrac{\pi}{3}\right) =$

    (A) $\sqrt{3}$

    (B) $-\dfrac{\sqrt{3}}{2}$

    (C) 0

    (D) $\dfrac{\sqrt{3}}{2}$

    (E) $\sqrt{3}$

74. What is the slope of the tangent to the curve $y^2(x^2 + y^2) = 3x^2$ at $(2, \sqrt{2})$?

    (A) $-2\sqrt{3}$

    (B) $\sqrt{2}$

    (C) $\dfrac{\sqrt{2}}{2}$

    (D) $\dfrac{\sqrt{2}}{4}$

    (E) $\dfrac{\sqrt{2}}{8}$

75. What is the slope of the tangent to the curve $\sin(\pi x) + 9\cos(\pi y) = x^2 y$ at $(3, -1)$?

    (A) $\dfrac{\pi - 9}{6}$

    (B) $\dfrac{6}{\pi - 9}$

    (C) $\dfrac{6 - \pi}{9}$

    (D) $\dfrac{9}{6 - \pi}$

    (E) $\dfrac{\pi}{6}$

76. Find $\dfrac{dy}{dx}$ if $x^2 y^2 - 3x = 5$.

(A) $\dfrac{2xy^2 - 3}{2x^2 y}$

(B) $\dfrac{3 - 2xy^2}{2x^2 y}$

(C) $\dfrac{2 - 3xy^2}{2x^2 y}$

(D) $\dfrac{2x^2 y - 3}{2x^2 y}$

(E) $\dfrac{2 - 3xy^2}{3x^2 y}$

77. If $f(x) = \dfrac{\sqrt{x}}{2}$ and $g(x) = \cos x$, find $[f(g(x))]'$.

(A) $\dfrac{\sin x}{4\sqrt{\cos x}}$

(B) $-\dfrac{\sin x}{4\sqrt{\cos x}}$

(C) $-\dfrac{\sin x}{2\sqrt{\cos x}}$

(D) $\dfrac{\sin x\sqrt{\cos x}}{4}$

(E) $\dfrac{\sin x\sqrt{\cos x}}{2}$

78. Given the table below, find the slope of the tangent line of $\dfrac{f(x)}{g(x)}$ at the point $x = 2$.

f(2)	g(2)	f'(2)	g'(2)
10	3	5	7

(A) $\dfrac{5}{7}$

(B) $\dfrac{19}{9}$

(C) $\dfrac{85}{9}$

(D) $-\dfrac{55}{9}$

(E) Undefined

79. Given the table below, find $h''(3)$, where $h(x) = f(x)g(x)$.

f(3)	g(3)	f'(3)	g'(3)	f''(3)	g''(3)
−1	2	−5	1	1	4

(A) −2
(B) −11
(C) −12
(D) −19
(E) −20

80. For $y = -\dfrac{1}{4}\log_2(5x^2 - 9)$, find $\dfrac{dy}{dx}$.

(A) $-\dfrac{10x}{4\ln 2(5x^2 - 9)}$

(B) $-\dfrac{10x}{\ln 2(5x^2 - 9)}$

(C) $-\dfrac{1}{4(5x^2 - 9)}$

(D) $-\dfrac{1}{4\ln 2(5x^2 - 9)}$

(E) $-\dfrac{10x}{4(5x^2 - 9)}$

81. For $f(x) = \sqrt{x} \log_4\left(\dfrac{1}{x}\right)$, find $f'(x)$.

(A) $\dfrac{\log_4\left(\dfrac{1}{x}\right) - 2}{2\sqrt{x}}$

(B) $\dfrac{\log_4\left(\dfrac{1}{x}\right) - 2}{\sqrt{x}}$

(C) $\dfrac{(\ln 4)\log_4\left(\dfrac{1}{x}\right) - 2}{2(\ln 4)\sqrt{x}}$

(D) $\dfrac{x}{2(\ln 4)\sqrt{x}}$

(E) $-\dfrac{\sqrt{x}}{2\ln 4}$

82. Find the slope of the tangent line to $f(x) = \dfrac{e^{\sqrt{x}}}{x}$ at the point $x = 4$.

(A) $-\dfrac{15e^2}{64}$

(B) $-\dfrac{3e^2}{64}$

(C) $\dfrac{5e^2}{64}$

(D) $-5e^2$

(E) 0

83. For $f(x) = \dfrac{\ln 2}{2^x}$, find $f'(x)$.

(A) $\dfrac{1}{2^x}$

(B) $-\dfrac{\ln 2}{4^x}$

(C) $\dfrac{-\ln^2(2)}{2^x}$

(D) $\dfrac{1-\ln(2)}{2^x}$

(E) $\dfrac{1}{2}\left(\dfrac{1-\ln^2(2)}{2^x}\right)$

84. For $f(x) = e^4 \cos^2(5x)$, find $\dfrac{dy}{dx}$.

(A) $-2e^4 \sin(5x)$

(B) $-2e^4 \sin(10x)$

(C) $-5e^4 \sin(10x)$

(D) $e^4(\cos^2(5x) - \sin(10x))$

(E) $e^4(\cos^2(5x) - 2\sin(5x))$

85. Suppose for a differentiable function f, $f(0) = 2$ and $f'(0) = 9$. Find $h'(0)$, where $h(x) = e^x f(x)$.

(A) 0
(B) 1
(C) 9
(D) 11
(E) Undefined

86. At $x = 3$, a function f has a value of 2 and a horizontal tangent line. If $h(x) = (f(x))^3$, find $h'(3)$.

(A) 0
(B) 4
(C) 6
(D) 12
(E) Undefined

87. For $y = \sin(x^3)$, find $\dfrac{d^2 y}{dx^2}$.

    (A) $-3x^2(\sin(x^3) - \cos(x^3))$

    (B) $3x(2\cos(x^3) - 3x^3 \sin(x^3))$

    (C) $9x^4 \cos^2(x^3)$

    (D) $3x^2 \cos(x^3)$

    (E) $-6x\sin(x^3)$

88. Given that $f(x) = 3x^3 + 5$, find the slope of the tangent line to $f^{-1}(x)$ at $x = 2$.

    (A) $-\dfrac{36}{841}$

    (B) $-\dfrac{1}{29}$

    (C) $\dfrac{1}{36}$

    (D) $\dfrac{1}{9}$

    (E) 36

89. If $f(3) = 8$, then the slope of the tangent line to $\left(f^{-1}\right)'(8)$ must be

    (A) $\dfrac{1}{f'(3)}$

    (B) $\dfrac{1}{f'(8)}$

    (C) $f'(3)$

    (D) 3

    (E) 8

BC 90. Evaluate $\lim\limits_{x \to 0} \dfrac{e^{2x} - 1}{\sin x}$.

    (A) 0

    (B) 1

    (C) 2

    (D) 4

    (E) Undefined

91. For a constant $n > 1$, find a general formula for $\dfrac{d}{dx}\left(\dfrac{1}{3}\sqrt{x}\right)^n$.

(A) $\dfrac{n}{6}x^{\frac{n-2}{2}}$

(B) $\left(\dfrac{n}{6}\right)^{n-1}x^{\frac{n-2}{2}}$

(C) $\left(\dfrac{n}{6}\right)^{n}x^{\frac{n-2}{2}}$

(D) $\left(\dfrac{1}{3}\right)^{n-1}\left(\dfrac{n}{2}\right)x^{\frac{n-2}{2}}$

(E) $\left(\dfrac{1}{3}\right)^{n}\left(\dfrac{n}{2}\right)x^{\frac{n-2}{2}}$

BC 92. Suppose that $f(0) = g(0) = 0$ for two differentiable functions f and g. The limit of $\dfrac{f}{g}$ as x approaches zero is equivalent to

(A) The limit of $\dfrac{f'}{g'}$ as x approaches zero.

(B) The limit of $\dfrac{f'g - fg'}{g^2}$ as x approaches zero.

(C) The limit of $\dfrac{fg' - f'g}{g^2}$ as x approaches zero.

(D) The limit of $\dfrac{f'}{g}$ as x approaches zero.

(E) The limit of $\dfrac{f}{g'}$ as x approaches zero.

93. The first derivative of a function is linear. Which of the following must be true of the second derivative of this function?

(A) It must be a positive constant.
(B) It must be a negative constant.
(C) It must be zero.
(D) It must be a constant which may be either positive or negative.
(E) Cannot be determined.

94. At which values of x does $\dfrac{\sec^2(x)}{4}$ have a horizontal tangent line?

(A) 0 only

(B) π only

(C) $n\pi$ for any integer n

(D) $\dfrac{n\pi}{2}$ for any integer n

(E) The function does not have any horizontal tangent lines

95. For a function f, the equation of the tangent line at $x = 2$ is $5x + 112$. For a function g, the equation of the tangent line at $x = 2$ is $5x + 83$. Which of the following statements must be true?

(A) $f'(2) - g'(2) = 29$

(B) $f(2) = g(2)$

(C) $f'(2) = g'(2)$

(D) $f(2) - g(2) = 29$

(E) No conclusion can be made without knowing the exact form of f and g.

96. Let f be the function given by $f(x) = \dfrac{\sin^2(x)}{x}$.

(A) Find the derivative of $f(x)$.

(B) Write an equation for the tangent line at $x = \dfrac{\pi}{2}$.

97. Let f, g, and h be differentiable functions.

(A) Find a general formula for $\left(\dfrac{fg}{h} \right)'$.

(B) Use your formula from part (a) to find $\left(\dfrac{e^x \sin(x)}{x^2} \right)'$.

98. Suppose that f is a differentiable function. Find the derivative of $h(x) = \dfrac{(f(x))^4}{\sqrt[3]{x}}$.

99. Find the equation of the tangent line of $f(x) = \dfrac{x^2}{\cos x}$ at $x = \dfrac{\pi}{4}$.

100. Are there any points where both $f(x) = \dfrac{3x^3 + 5}{\ln x}$ and $g(x) = 4x - 9$ have horizontal tangent lines? Justify your answer.

Graphs of Functions and Derivatives

101. The derivative of g is given by $g'(x) = x^2 \sin(x - 1)$. Where on the interval $-\pi < x \le \pi$ does the function g have a relative maximum?

(A) $x = 1 - \pi$
(B) $x = 0$
(C) $x = 1$
(D) $x = 0$ and $x = \pi$
(E) $x = 1 - \pi$ and $x = 1$

102. What are all values of x for which the function f defined by $f(x) = xe^{-(x-7)}$ is increasing?

(A) $x < -7$
(B) $x < 1$
(C) $x > 1$
(D) $-7 < x < 1$
(E) $x < -7$ and $x > 1$

103. At what value of x does the function $y = 1.2x^2 - e^{4x}$ change concavity?

(A) $x = 2.5 \ln 6$
(B) $x = 6 \ln 2.4$
(C) $x = 2.4 \ln 12$
(D) $x = 2.5 \ln 15$
(E) $x = 15 \ln 1.6$

104. Let f be the function given by $f(x) = \dfrac{x}{x - 3}$. For what value(s) of x is the slope of the line tangent to the graph of f at $(x, f(x))$ equal to $-\dfrac{3}{4}$?

(A) $x = -\dfrac{3}{7}$

(B) $x = \dfrac{9}{7}$

(C) $x = 1$ or $x = 5$
(D) $x = \pm 7$
(E) No solution

105. The graph of the derivative of $h(x)$ is given.

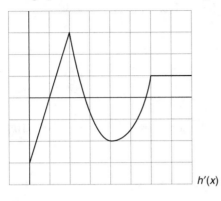

$h'(x)$

Which of the following could be the graph of $h(x)$?

(A)
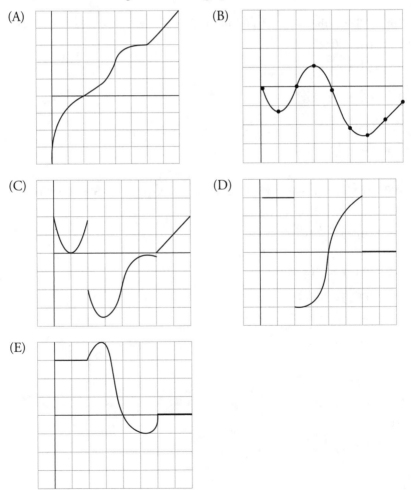

(B)

(C)

(D)

(E)

106. According to the mean value theorem, there exists at least one $x = c$ on the interval $1 < x < b$ such that $f'(c) = -1/2$. Find b if $f(x) = 8x - x^3$.

(A) $b = 1.68$
(B) $b = 2.19$
(C) $b = 2.28$
(D) $b = 2.82$
(E) $b = 2.89$

107. Which of the following is an inflection point of $p(x) = x^4 - 2x^3 + 10x - 8$?

(A) $(-8, 0)$
(B) $(-1, -15)$
(C) $(0, 8)$
(D) $(1, 1)$
(E) $(12, -1)$

108. What is the rate of change of the function $f(x) = \sqrt{9 - x^3}$ at $x = 2$?

(A) -6
(B) $-\dfrac{3}{2}$
(C) $-\dfrac{1}{2}$
(D) 1
(E) $\dfrac{1}{2}$

109. The line tangent to the function $g(x) = \ln(x^2 + x + 6)$ at $x = 0$ is

(A) $y = x \ln 6$
(B) $y = \dfrac{1}{6} \ln 6$
(C) $y = \dfrac{1}{6}(x - \ln 6)$
(D) $y = 6x - \ln 6$
(E) $y = \dfrac{1}{6}x + \ln 6$

110. The graph of a function $g(x)$ is given.

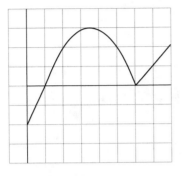

Which of the following could be the graph of $g'(x)$?

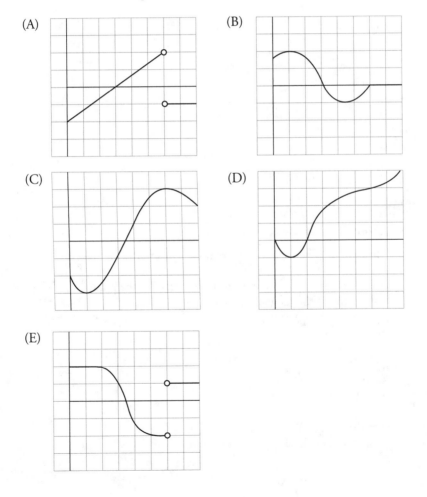

111. For which values of t is the function in the xy plane defined as $x = 4 - t$ and $y = t^2 + t$ increasing?

(A) $t < -\dfrac{1}{2}$

(B) $t < 0$

(C) $t > \dfrac{1}{2}$

(D) $t > \dfrac{9}{2}$

(E) $t < 4$

112. If $y = \ln(8 - x^3)$, then $\dfrac{dy}{dx} =$

(A) $\dfrac{3x^2}{x^3 - 8}$

(B) $\dfrac{3x^2}{8 - x^3}$

(C) $\dfrac{-3}{8 - x}$

(D) $\dfrac{1}{8 - x^3}$

(E) $\dfrac{\ln(8 - x^3)}{-3x^2}$

113. A particle travels along the curve defined by $x = \cos(2t)$, $y = \sin(t)$ starting when $t = 0$. When $t = \dfrac{\pi}{4}$, the particle stops following the curve and continues along the line tangent to the curve at the point $\left(0, \dfrac{\sqrt{2}}{2}\right)$. What is the slope of the tangent line?

(A) -2

(B) -1

(C) $\dfrac{-\sqrt{2}}{4}$

(D) 0

(E) $\dfrac{\sqrt{2}}{2}$

114. The function f is continuous and differentiable on $0 \le x \le 10$. Use the table of values to determine an interval for which according to Rolle's theorem $f'(c) = 0$ for some c on the interval.

X	f(x)
0	2
1	5
3	6
6	7
7	5
9	4
10	1

(A) $6 < x < 7$
(B) $5 < x < 7$
(C) $1 < x < 7$
(D) $5 < x < 6$
(E) $3 < x < 6$

115. The table below includes all critical points of the continuous function $g(x)$. Use the table to determine where the function $g(x)$ is increasing within the interval $-2 < x < 20$.

X	g(x)
-2	1.47
-0.94	0
0	-0.33
9.08	0
14	-7.3
20	-4.40

(A) $0 < x < 20$
(B) $0 < x < 9.08$ and $14 < x < 20$
(C) $-2 < x < 0$
(D) $-2 < x < -.94$
(E) $-2 < x < -.94$ and $14 < x < 20$

116. The table below gives specific values of the derivative of the function $h(x)$ and includes all critical points of the function $h(x)$. What is the maximum value of $h(x)$?

X	h′(x)
−10	33.9
−2.2	0
0	−9.9
3.7	0
5	−2
9.3	Does not exist
10	8

(A) $h(-10)$
(B) $h(-2.2)$
(C) $h(3.7)$
(D) 33.9
(E) $h(x)$ may or may not have a maximum value.

117. The second derivative of $f(x)$ has zeros at $x = a$ and $x = c$ and a minimum at $x = b$ as shown. The function $f(x)$ is concave up

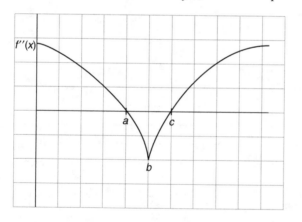

(A) when $0x < a$
(B) when $0x < b$
(C) when $x > b$
(D) when $0x < a$ and $x > c$
(E) nowhere

118. If the line tangent to $y = f(x)$ at point $(-3, 8)$ passes through the point $(-2, 5)$ then,

(A) $f'(-2) = 3$

(B) $f'(-2) = -3$

(C) $f'(-3) = \dfrac{3}{5}$

(D) $f'(-3) = -3$

(E) $f(-2) = 5$

119. The derivative of f has a zero at $x = a$ and a relative maximum at $x = b$, as shown. Which of the following is not true?

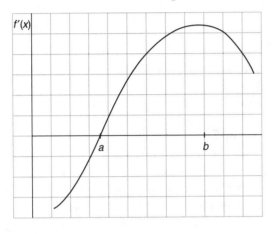

(A) $f(x)$ has a relative minimum at $x = a$.

(B) $f(x)$ has an absolute maximum at $x = b$.

(C) $f(x)$ is increasing on (a, b).

(D) $f''(x)$ is positive on (a, b).

(E) $f''(x)$ has a zero at $x = b$.

120. The graph of $g(x)$ has zeros at $x = k$ and $x = n$ and a relative maximum at m as shown. Based on the graph, which of the following is true?

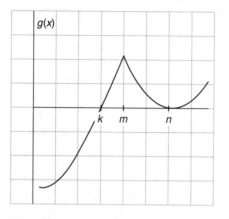

(A) $g'(x)$ has a relative maximum at $x = k$.
(B) $g'(x)$ has a zero at $x = m$.
(C) $g''(x)$ has a zero at $x = n$.
(D) $g'(x)$ is continuous everywhere.
(E) $g''(x)$ is never negative.

121. Let $f(x) = x^3 - x$ such that f is continuous on a closed interval $[-1, 1]$. Find the critical number(s), c, that satisfies the mean value theorem for the given function and interval.

(A) $\pm\sqrt{\dfrac{1}{3}}$

(B) $\pm\sqrt{3}$

(C) 0

(D) $\sqrt[3]{\dfrac{1}{3}}$

(E) $\dfrac{1}{3}$

BC 122. Approximate the angle between vectors $r_1 = \langle 2, 3 \rangle$ and $r_2 = \langle 6, 4 \rangle$ in radians.

(A) 0.040 radians
(B) 0.281 radians
(C) 0.395 radians
(D) 1.017 radians
(E) 1.571 radians

123. Find the extrema of function $f(x) = \frac{1}{3}x^3 - 6x^2 + 35x - 1$.

    (A) absolute minimum at $x = 0$
    (B) absolute maxima at $x = 5, 7$
    (C) relative maximum at $x = -5$, relative minimum at $x = -7$
    (D) relative maximum at $x = 5$, relative minimum at $x = 7$
    (E) relative maximum at $x = 7$, relative minimum at $x = 5$

124. Find all critical points, c, for the function $f(x) = \frac{2}{3}x^3 + 5x^2 - 28x - 10$.

    (A) $c = 0, -7, -2$
    (B) $c = -7, 2$
    (C) $c = 0$
    (D) $c = -2, 7$
    (E) $c = 10$

125. Find the inflection point(s) for the function $f(x) = 2x(x + 4)^3$.

    (A) $(0, 0)$
    (B) $(-4, 0)$
    (C) $(0, 0), (-4, 0)$
    (D) $(-4, 0), (0, 0), (4, 0)$
    (E) $(-4, 8)$

BC 126. Find the polar equation of the ellipse $\dfrac{x^2}{25} + \dfrac{y^2}{16} = 1$.

    (A) $r = 16 \cos^2\theta + 25 \sin^2\theta$

    (B) $r = \dfrac{\cos^2\theta}{25} + \dfrac{\sin^2\theta}{16}$

    (C) $r = \dfrac{20}{4\cos\theta + 5\sin\theta}$

    (D) $r = \dfrac{20}{\sqrt{16 + 9\sin^2\theta}}$

    (E) $r = \dfrac{20}{4 + 3\sin\theta}$

127. Let $f(x) = \sin x$ on the interval $\left[0, \dfrac{\pi}{2}\right]$. Find an approximation to the number(s) c that satisfies the mean value theorem for the given function and interval.

    (A) 0.3014
    (B) 0.4404
    (C) 0.6366
    (D) 0.8041
    (E) 0.8807

128. Which of the following statements is true of the function $(x) = x^{\frac{2}{3}}$?
  I. There is a critical point at $(0, 0)$.
  II. $f'(0)$ and $f''(0)$ are undefined.
  III. The curve is concave up over the interval $(0, +\infty)$.
  IV. The curve is concave down over interval $(-\infty, 0)$.
  (A) I and III only
  (B) I, II, and IV only
  (C) I, II, and III
  (D) I, III, and IV
  (E) I, II, III, and IV

129. If $f'(x) = 2x^2 - 5$, find the interval(s) where f is decreasing.

  (A) $\left(-\infty, \ -\sqrt{\dfrac{5}{2}}\right)$

  (B) $\left(-\infty, \ \sqrt{\dfrac{5}{2}}\right)$

  (C) $\left(-\sqrt{\dfrac{5}{2}}, \ +\infty\right)$

  (D) $\left(-\sqrt{\dfrac{5}{2}}, \ \sqrt{\dfrac{5}{2}}\right)$

  (E) $\left(\sqrt{\dfrac{5}{2}}, \ +\infty\right)$

BC 130. Find the components of the vector of magnitude 6 and direction $\dfrac{\pi}{6}$.
  (A) $\langle 3\sqrt{3}, \ 3\rangle$
  (B) $\langle \sqrt{3}, \ 0\rangle$
  (C) $\langle 2\sqrt{3}, \ 3\rangle$
  (D) $\langle 1, \ \sqrt{3}\rangle$
  (E) $\langle \pi, \ \pi\rangle$

131. A function $f(x) = \dfrac{1}{x-1}$. Determine the concavity for intervals $(-\infty, 1)$ and $(1, +\infty)$, respectively.
  (A) concave up, concave up
  (B) concave down, concave up
  (C) concave down, concave down
  (D) concave up, concave down
  (E) concave up, no concavity

132. Determine the symmetry, if any, of the graph of $r = 6\cos(3\theta)$

  (A) symmetric about the x-axis, the y-axis, and the pole
  (B) symmetric about the x-axis and the pole
  (C) symmetric about the pole and the y-axis
  (D) symmetric about the pole only
  (E) symmetric about the x-axis only

133. Determine the intervals on which $f(x) = x^3 - x^2$ increases and the intervals on which it decreases.

  (A) increasing on $(-\infty, 0)$, decreasing on $(0, +\infty)$
  (B) decreasing on $(-\infty, 0)$, increasing on $(0, +\infty)$
  (C) increasing on $(-\infty, 0) \cup \left(\frac{2}{3}, +\infty\right)$, decreasing on $\left(0, \frac{2}{3}\right)$
  (D) decreasing on $(-\infty, 0) \cup \left(\frac{2}{3}, +\infty\right)$, increasing on $\left(0, \frac{2}{3}\right)$
  (E) increasing on $\left(-\infty, \frac{2}{3}\right)$, decreasing on $\left(\frac{2}{3}, +\infty\right)$

134. If $f(x) = |x^2 - 4|$, which of the following statements about f are true?

  I. f is continuous on the interval $(-\infty, +\infty)$.
  II. f has points of inflection at $x = \pm 2$.
  III. f has a relative maximum at $(0, 4)$.

  (A) I only
  (B) II only
  (C) III only
  (D) II and III
  (E) I and III

135. Find the length and direction of the vector represented by $\langle 3, 3\sqrt{3}\rangle$.

  (A) $\|r\| = 3\sqrt{10}$, $\theta = \frac{\pi}{3}$
  (B) $\|r\| = 3\sqrt{10}$, $\theta = \frac{\pi}{6}$
  (C) $\|r\| = 6$, $\theta = \frac{\pi}{3}$
  (D) $\|r\| = 6$, $\theta = \frac{2\pi}{3}$
  (E) $\|r\| = 6$, $\theta = \frac{\pi}{6}$

136. Find the critical point(s) of the function $f(x) = 4x^2 - 3x + 2$.

(A) $-\dfrac{1}{4}, -\dfrac{1}{2}$

(B) $-\dfrac{3}{8}$

(C) 0

(D) $\dfrac{3}{8}$

(E) $\dfrac{1}{4}, \dfrac{1}{2}$

137. Determine the values of a and b in the function $f(x) = ax^3 + b\left(\dfrac{1}{x}\right)$ such that f has a minimum at $(1, 4)$ and maximum at $(-1, -4)$.

(A) $a = -2, b = 6$

(B) $a = \dfrac{1}{3}, b = 1$

(C) $a = 2, b = 2$

(D) $a = 1, b = 3$

(E) $a = 3, b = 1$

BC 138. Find the equation for $r^2 \cos 2\theta = 1$ in Cartesian coordinates.

(A) $\dfrac{x^2}{2} + \dfrac{y^2}{2} = 1$

(B) $(x - 2)^2 + y^2 = 1$

(C) $(x + 2)^2 - y^2 = 1$

(D) $x^2 + y^2 = 1$

(E) $x^2 - y^2 = 1$

139. Verify whether $f(x) = 3x^2 - 12x + 1$ satisfies Rolle's theorem on the interval $[0, 4]$ and find all numbers c that satisfy $f'(c) = 0$.

(A) $c = 0$

(B) $c = 1$

(C) $c = 2$

(D) $c = 4$

(E) $f(x)$ does not satisfy Rolle's theorem on interval $[0, 4]$.

140. Find the interval(s) where $f(x)$ is increasing if $f'(x) = x^4 - 16$.

(A) $(-2, 2)$

(B) $(0, 2)$

(C) $(-\infty, -2) \cup (2, +\infty)$

(D) $(-2, 0) \cup (2, +\infty)$

(E) $(-\infty, -2) \cup (0, 2)$

BC 141. A parametric curve is given by $x = \ln t$ and $y = 4t + 1$. Find the Cartesian equation of the curve.

(A) $y = -\dfrac{1}{4 \ln x}$

(B) $y = \dfrac{1}{4 \ln x}$

(C) $y = 4 \ln x + 1$

(D) $y = 4e^x + 1$

(E) $y = \ln x$

142. Find the relative maximum and minimum values for the function $f(x) = 2x^3 + x^2 + 15$ on the interval $[-4, 4]$.

(A) minimum: $f(0) = 15$, maximum: $f(3) = 78$

(B) minimum: $f(2) = 35$, maximum: $f(-2) = 3$

(C) minimum: $f\left(-\dfrac{1}{3}\right) = \dfrac{406}{27}$, maximum: $f(0) = 15$

(D) minimum: $f(0) = 15$, maximum: $f\left(-\dfrac{1}{3}\right) = \dfrac{406}{27}$

(E) minimum: $f(-2) = 3$, maximum: $f(2) = 3$

143. $f'(x) = 3x^2 - 2x + 1$. Find the concavity of function f on the intervals $\left(-\infty, \dfrac{1}{3}\right)$ and $\left(\dfrac{1}{3}, +\infty\right)$, respectively.

(A) concave down, concave up

(B) concave up, concave down

(C) concave down, concave down

(D) concave up, concave up

(E) concave down, no concavity

144. What shape is described by the equation $r = 5 \cos 4\theta$?

(A) limacon with inner loop

(B) limacon with no inner loop

(C) rose with 8 petals of length 5

(D) rose with 4 petals of length 5

(E) cardioid

145. Which of the following statements are true of the graph of f below?

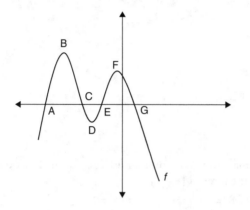

   I. $f' \geq 0$ on the interval from D to F
  II. $f'' = 0$ at points B, D, and F
 III. $f'' > 0$ on the interval from A to B
 IV. $f'' > 0$ on the interval from D to F

  (A) I and II
  (B) I and III
  (C) II and IV
  (D) II, III, and IV
  (E) I, II, and III

146. $f(x) = 1 - \sqrt[3]{x}$
  (A) Find the intervals on which f is increasing or decreasing.
  (B) Locate all maxima and minima.
  (C) Find the intervals over which f is concave upward or downward.
  (D) Find all inflection points.
  (E) Sketch the graph of f.

147. There are two vectors $(1, -4)$ and $\langle 2, k \rangle$, where k is an unknown quantity.
  (A) Find a value of k such that the vectors are orthogonal.
  (B) Find a value of k such that the vectors are parallel.
  (C) If $k = 6$, find the angle between the two vectors. Round to the
     nearest tenth of a degree.

BC 148. A force of 2 newtons is applied to the right side of an object. A force of 4 newtons is applied from below. A force of 8 newtons is applied to the left side of the object at an angle of $\dfrac{\pi}{3}$ above the horizontal.

(A) Find the resultant force vector being applied to the object.
(B) Find an approximation (to the nearest hundredth) of the magnitude of the total force on the object.
(C) What additional force vector would need to be applied to keep the object from moving?

149. Given the graph of f', find the following properties of the function f:

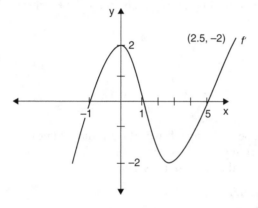

(A) The intervals on which f is increasing or decreasing
(B) The location of the relative maxima and minima
(C) The points of inflection and concavity of f
(D) Draw a sketch of f, given that $f(-1) = f(1) = 5, f(0) = 0$ and $f(5) = -5$.

150. $f(x) = x^4 - x^3$.

(A) Find the intervals on which f is increasing or decreasing.
(B) Locate all maxima and minima
(C) Find the points of inflection, if any, on f.
(D) Find the intervals where f is concave upward or downward.
(E) Sketch the graph of $f(x) = x^4 - x^3$.

4 CHAPTER

Applications of Derivatives

151. The position of a moving car is given by $f(x) = 2x + 5$ meters. What is the velocity of the car?

(A) 0.5 m/s
(B) 1.0 m/s
(C) 2.0 m/s
(D) 2.5 m/s
(E) 5.0 m/s

152. A spherical balloon is being inflated. What is the volume of the sphere at the instant when the rate of increase of the surface area is four times the rate of increase of the radius of the sphere?

(A) $\dfrac{1}{2\pi}$ cubic units

(B) $\dfrac{3\pi r^2}{2}$ cubic units

(C) $4\pi^2$ cubic units

(D) $\dfrac{1}{6\pi^2}$ cubic units

(E) $8\pi r$ cubic units

153. A conical funnel has a base diameter of 4 cm and a height of 5 cm. The funnel is initially full, but water is draining at a constant rate of 2 cm³/s. How fast is the water level falling when the water is 2.5 cm high?

(A) $\dfrac{1}{2\pi}$ cm/s

(B) $-\dfrac{1}{2\pi}$ cm/s

(C) $\dfrac{2}{\pi}$ cm/s

(D) $-\dfrac{2}{\pi}$ cm/s

(E) -2π cm/s

154. A light on the ground 50 m from a tall building shines on a 1-m tall child. The child is walking away from the light at a velocity of 1 m/s. How fast is the shadow on the building decreasing when the child is 10 m away from the building?

(A) $-\dfrac{5}{4}$ m/s

(B) $-\dfrac{1}{50}$ m/s

(C) $-\dfrac{1}{32}$ m/s

(D) $-\dfrac{1}{20}$ m/s

(E) $-\dfrac{4}{5}$ m/s

155. A camera on the ground is 100 m away from a model rocket launch pad. When the rocket is launched, it rises at a constant velocity of 20 m/s. How fast must the camera's angle change when the rocket is 100 m high?

(A) 0.57 deg/s
(B) 1.1 deg/s
(C) 2.3 deg/s
(D) 5.7 deg/s
(E) 11.4 deg/s

156. The width of a rectangle increases twice as fast as the length. How does the rate of change of the area of the rectangle compare to that of the rate of change of the width when the length of the rectangle is twice the width?

(A) w
(B) w/2
(C) 2w
(D) 2.5w
(E) 5w

157. Given the cost function $C(x) = 200 + 4x + 0.5x^2$, what is the marginal cost, when $x = 50$?

(A) $5
(B) $54
(C) $200
(D) $250
(E) $1650

158. The position of a particle with time (seconds) can be described by the following function: $f(x) = x^3 - 6x^2 + 9x$. At what times, will the velocity of the particle be zero?

(A) 0 and 4 s
(B) 1 and 3 s
(C) 0 and 3 s
(D) 1 and 2 s
(E) 2 and 4 s

159. You have 500 ft of fence to enclose your dog in a rectangular area. What are the dimensions of the fence that will give you the maximum area?

(A) length = 50 ft, width = 200 ft
(B) length = 100 ft, width = 150 ft
(C) length = 75 ft, width = 175 ft
(D) length = 180 ft, width = 70 ft
(E) length = 125 ft, width = 125 ft

160. A man and a woman leave the same intersection and walk away at the same speed. The man walks due east at 1 m/s, while the woman walks due south at 1 m/s. When the man and the woman are each 100 m away from the intersection, how fast is the distance between them increasing?

(A) 1.0 m/s
(B) 1.4 m/s
(C) 2.0 m/s
(D) 2.8 m/s
(E) 3.0 m/s

161. What is the instantaneous rate of change of this function: $f(x) = 2x^2 - 4x + 2$, when $x = 20$?

(A) 20
(B) 76
(C) 152
(D) 3042
(E) 6396

162. The position in meters of a moving object can be described by this function: $f(x) = 2x^3 - 4x^2 + 2x - 5$. What is the instantaneous acceleration of the object at $x = 10$?

(A) 112 m/s^2
(B) 120 m/s^2
(C) 522 m/s^2
(D) 1120 m/s^2
(E) 1200 m/s^2

163. The cost of producing a brand of computer is given by this function $C(x) = 200 + 16x + 0.1x^2$. If the computers sell for $500 each and 1000 are produced and sold, then what is the marginal profit?

(A) $100
(B) $142
(C) $200
(D) $284
(E) $500

164. A right cylindrical cone has a radius of 4 cm and a height of 2.0 cm. If the height increases at 0.5 cm/min, but the radius remains constant, then what will be the rate of change of volume?

(A) 1.1 cm³/min
(B) 2.1 cm³/min
(C) 4.2 cm³/min
(D) 6.3 cm³/min
(E) 8.4 cm³/min

165. What are the coordinates of the vertex of the parabola $y = 3x^2 + 2$?

(A) (0, 0)
(B) (−2, 0)
(C) (0, −2)
(D) (0, 2)
(E) (2, 0)

BC 166. Water is continuously pumped into and removed from a cylindrical holding tank with a radius of 10 m. If the water is pumped in at the rate of 100 m³/min and removed at the rate of 70 m³/min, then what is the approximate rate of change of the water level in the tank?

(A) −1.0 m/min
(B) −0.1 m/min
(C) 0 m/min
(D) 0.1 m/min
(E) 1.0 m/min

167. An 8-ft ladder is leaning against a wall. If the top of the ladder is sliding down the wall at 1 ft/s, how fast is the bottom of the ladder sliding away from the wall when the top is 4 ft from the ground?

(A) $\sqrt{3}$ ft/s

(B) $\sqrt{2}$ ft/s

(C) $\dfrac{\sqrt{3}}{3}$ ft/s

(D) $\dfrac{\sqrt{3}}{2}$ ft/s

(E) $\dfrac{\sqrt{2}}{2}$ ft/s

168. A 1.0-kg iron block is heated. The amount of heat stored in the cylinder is given by $Q = mc\Delta T$, where Q is the thermal energy (J), m is the mass in kg, c is the specific heat capacity (c of iron = 460 J/kg·K), and ΔT is the change in temperature (kelvin, K). If the block is heated so that the change in thermal energy is 100 J/s, then what is the approximate rate at which the temperature changes?

(A) 0.1 K/s
(B) 0.22 K/s
(C) 0.50 K/s
(D) 0.8 K/s
(E) 1.0 K/s

169. A spherical balloon is being inflated at a rate of 50 cm³/s. How fast is the diameter increasing when the radius of the balloon is 10 cm?

(A) $\dfrac{1}{\pi}$ cm/s

(B) $\dfrac{2}{\pi}$ cm/s

(C) $\dfrac{1}{2\pi}$ cm/s

(D) $\dfrac{1}{3\pi}$ cm/s

(E) $\dfrac{1}{4\pi}$ cm/s

170. A farmer has 160 m of fence to enclose a rectangular area against a straight river. He only needs to fence in three sides. What is the maximum area that he can enclose with his materials?

(A) 160 m²
(B) 320 m²
(C) 960 m²
(D) 1920 m²
(E) 3200 m²

171. Given the cost function $C(x) = 144 + 0.1x + 0.04x^2$, what is the minimum average cost per unit?

(A) $20
(B) $40
(C) $60
(D) $80
(E) $100

172. A graph of $y = -2x + 4$ encloses a region of the x and y axes in the first quadrant. An inscribed rectangle has one vertex at the origin and the opposite one on the graph. What are the dimensions of the rectangle that has the maximum area?

(A) length = 0.5, width = 2
(B) length = 1, width = 2
(C) length = 1, width = 1
(D) length = 2, width = 1
(E) length = 2, width = 2

173. Find two negative numbers that add up to -50 such that the maximum product is possible.

(A) $-5, -45$
(B) $-10, -40$
(C) $-15, -35$
(D) $-20, -30$
(E) $-25, -25$

174. A spherical balloon is being inflated. At some point, the volume of the sphere is $1/(48\pi^2)$ cubic units. At that instant, how does the rate of increase of the surface area compare to the rate of increase of the radius of the sphere?

(A) They are equal.
(B) The increase in surface area is one-half as fast as the rate of increase in the radius.
(C) The rate of increase is one-fourth as fast as the rate of increase in the radius.
(D) The rate of increase of the surface area is twice that of the radius.
(E) The rate of increase of the surface area is four times that of the radius.

175. You want to make an open box from a 4 cm by 8 cm sheet of cardboard by cutting a square from each corner and folding the sides up. What is the length of the square that will yield the greatest volume?

(A) 0.1 cm
(B) 0.5 cm
(C) 0.85 cm
(D) 1.7 cm
(E) 1.9 cm

176. Given the cost function $C(x) = 500 + 6x + 0.2x^2$ and the price function $p(x) = 20$, how many units should be produced to have the maximum profit?

(A) 10
(B) 14
(C) 20
(D) 35
(E) 50

177. Find two positive numbers that add up to 30 such that the maximum product is possible.

(A) 3, 27
(B) 5, 25
(C) 10, 20
(D) 15, 15
(E) 21, 29

178. Two cars leave the same intersection and drive away. Car A drives due east at 100 km/h, while car B drives due south at 50 km/h. After 1 h from the intersection, how fast is the distance between them increasing?

(A) 10 km/h
(B) 28 km/h
(C) 56 km/h
(D) 112 km/h
(E) 150 km/h

179. A 1.0-kg copper block is heated. The amount of heat stored in the cylinder is given by $Q = mc\Delta T$, where Q is the thermal energy (J), m is the mass in kg, c is the specific heat capacity (c of copper = 390 J/kg·K), and ΔT is the change in temperature (kelvin, K). If the block is heated so that the rate of change in temperature is 0.026 K/s, then what is the approximate rate at which the thermal energy was added?

(A) 1.0 J/s
(B) 2.0 J/s
(C) 5.0 J/s
(D) 8.0 J/s
(E) 10 J/s

180. A piano is suspended by a 90-ft rope through a pulley that is 40 ft above a man's arms. The piano is at some height above the ground. At $t = 0$, the man is 30 ft horizontally from the piano and walks away at 12 ft/s. How fast is the piano being pulled up?

(A) 1.0 ft/s
(B) 2.1 ft/s
(C) 4.5 ft/s
(D) 5.2 ft/s
(E) 10.4 ft/s

181. The position of a particle with time (seconds) can be described by the following function: $f(x) = x^3 - 9x^2 + 24x$. At what time, will the acceleration of the particle be zero?

(A) 0 s
(B) 1 s
(C) 2 s
(D) 3 s
(E) 4 s

182. A long, rectangular piece of metal, 16 in wide will be used to make a gutter. Two sides of equal length (x) will be turned up perpendicularly. How many inches should be turned up to give the maximum capacity?

(A) 1 in
(B) 2 in
(C) 3 in
(D) 4 in
(E) 7 in

183. An ellipsoid has the following radii: $r_a = 8$ cm, $r_b = 4$ cm, and $r_c = 2$ cm. r_a remains constant, but r_b increases by 0.5 cm/min and r_c increases by 2 cm/min. The equation for the volume of an ellipsoid is $V = (4/3)\pi r_a r_b r_c$. What will be the rate of change of volume?

(A) 13π cm^3/min
(B) 16π cm^3/min
(C) 64π cm^3/min
(D) 96π cm^3/min
(E) 128π cm^3/min

184. A plane lifts off from a runway at an angle of 30°. If the spe :d of the plane is 500 mi/h, then how fast is the plane gaining altitud :?

(A) 50 mi/h
(B) 100 mi/h
(C) 150 mi/h
(D) 200 mi/h
(E) 250 mi/h

185. The cost function of producing a product is $C(x) = 300 + 4x + 0.2x^2$. If the product sells for $500 each, then what is marginal profit function?

(A) $P'(x) = 4 + 0.4x$
(B) $P'(x) = 500 - 0.4x$
(C) $P'(x) = 504 + 0.4x$
(D) $P'(x) = 496 - 0.4x$
(E) $P'(x) = 496 + 0.4x$

186. A company orders 600 m of fence to enclose a rectangular area of their property against a straight river. The company only needs to fence in three sides. What is the maximum area that can be enclosed with these materials?

(A) 6000 m²
(B) 36000 m²
(C) 42000 m²
(D) 45000 m²
(E) 90000 m²

187. Oil is continuously pumped into and removed from a cylindrical holding tank with a radius of 100 m. If the oil is pumped in at the rate of 5000 m³/min and removed at the rate of 100 m³/min, then what is the approximate rate of change of the oil level in the tank?

(A) −16 m/min
(B) −0.16 m/min
(C) 0 m/min
(D) 0.16 m/min
(E) 16 m/min

188. The position in meters of a moving object can be described by this function: $f(x) = 16x^3 - 14x^2 + 6x - 7$. What is the instantaneous velocity of the object at $x = 0.5$?

(A) 1 m/s
(B) 2 m/s
(C) 4 m/s
(D) 8 m/s
(E) 16 m/s

189. What is the maximum volume of a cylinder that can be inscribed in a cone with an altitude of 15 cm and a base radius of 3 cm?

(A) π cm^3
(B) 2π cm^3
(C) 4π cm^3
(D) 16π cm^3
(E) 20π cm^3

190. The pressure across a bubble (Δp) is given by this equation: $\Delta p = 2\gamma/r$, where γ is the surface tension and r is the radius of the bubble. As the pressure across the bubble increases, the bubble shrinks. So, how does the rate of increase in pressure change compare to the rate of change of the radius when the rate of change of the surface area is four times the rate of decrease of the radius of the bubble (assume that γ is constant)?

(A) $-8\pi^2\gamma$ pressure units/time
(B) $-1/8\pi^2\gamma$ pressure units/time
(C) $-\pi^2/8\gamma$ pressure units/time
(D) $-8\gamma/\pi^2$ pressure units/time
(E) $-\pi^2/8\gamma$ pressure units/time

191. A tin can has a base diameter of 10 cm and a height of 10 cm. The can is initially full, but a hole is poked in the bottom center of the can. Water is draining at a constant rate of 4 cm^3/s. How fast is the water level falling?

(A) $\dfrac{1}{\pi}$ cm/s

(B) $-\dfrac{1}{2\pi}$ cm/s

(C) $-\dfrac{4}{25\pi}$ cm/s

(D) $-\pi$ cm/s
(E) -2π cm/s

192. Given the cost function $C(x) = x^2 + 20x + 4$, what is x so that the average cost function is minimum?

(A) 2
(B) 4
(C) 6
(D) 10
(E) 20

193. The velocity profile of a rocket can be described by the following function: $\frac{1}{3}t^3 - 2t^2 + 3t + 2$. At what time or times does the acceleration equal zero?

(A) 0 and 4 s
(B) 1 and 2 s
(C) 2 s
(D) 2 and 4 s
(E) 1 and 3 s

194. The height of a football when punted into the air is given by the function: $y(t) = v_o t - \frac{1}{2}gt^2$. The initial velocity of the football is v_o, g is acceleration due to gravity (10 m/s²), and t is time in seconds. If a football is kicked with an initial velocity of 20 m/s, then how long will it take to reach its maximum height?

(A) 0.5 s
(B) 1.0 s
(C) 1.5 s
(D) 2.0 s
(E) 2.5 s

195. You are given a position function for a moving object: $f(x) = 2x^2 - 20x + 5$ in meters. What is the object's acceleration?

(A) 0 m/s²
(B) 1 m/s²
(C) 2 m/s²
(D) 3 m/s²
(E) 4 m/s²

BC 196. A student builds an experimental model rocket with a variable thrust engine. The test flight lasted for 5 s and the rocket's altitude (in meters) could be described by the following function:

$$y(t) = \frac{1}{3}t^3 - 2t^2 + 3t + 2$$

(A) What is the velocity function of the rocket (m/s)? What is the acceleration function of the rocket (m/s²)?
(B) Make graphs of each function vs. time(s) for 0 to 5 s. Derive the maxima and minima for each graph (show your work).

197. A conical funnel has a base diameter of 6 cm and a height of 5 cm. The funnel sits over a cylindrical can with an open top. The can has a diameter of 4 cm and a height of 5 cm. The funnel is initially full, but water is draining from the funnel bottom into the can at a constant rate of 2 cm³/s. Answer the following questions: (show your work).

(A) How fast is the water level in the funnel falling when the water is 2.5 cm high?

(B) How fast is the water level in the can rising?

(C) Will the can overflow? If not, how high will the final level of water be in the can?

BC 198. The perimeter of an isosceles triangle is 16 cm. Do the following:

(A) Make a drawing of the problem.

(B) What are the dimensions of the sides and height for the maximum area?

(C) What is the maximum area?

199. A T-shirt maker estimates that the weekly cost of making x shirts is $C(x) = 50 + 2x + x^2/20$. The weekly revenue from selling x shirts is given by the function: $R(x) = 20x + x^2/200$. (Show your work).

(A) What is the profit if all the shirts made are sold?

(B) What is the maximum weekly profit?

200. You are to make a cylindrical tin can with closed top to hold 360 cm³.

(A) Make a drawing and label what is given.

(B) What are its dimensions if the amount of tin used is to be minimum?

(C) What is the surface area?

More Applications of Derivatives

201. A function y has a tangent at $x = a$ and a parallel tangent at $x = b$. Which of the following statements must be true?

(A) $\dfrac{dy}{dx}\big|_{x=a} = -\dfrac{dy}{dx}\big|_{x=b}$

(B) $\dfrac{dy}{dx}\big|_{x=a} = \dfrac{dy}{dx}\big|_{x=b}$

(C) $\dfrac{dy}{dx} = 0$

(D) $\dfrac{dy}{dx}\big|_{x=a} = \dfrac{dx}{dy}\big|_{x=a}$

(E) $\dfrac{dy}{dx}\big|_{x=a} \neq \dfrac{dy}{dx}\big|_{x=b}$

202. The velocity function of a moving particle is $v(t) = t^3 - 12t^2 + 5$. What is the particle's instantaneous velocity at $t = 3$?

(A) −81
(B) −76
(C) 76
(D) 81
(E) 94

203. What is the approximate value of $\sqrt[4]{78}$ using linear approximation?

(A) 2.64
(B) 2.68
(C) 2.75
(D) 2.86
(E) 2.97

BC 204. What is the slope of the line tangent to the curve defined by $r = 1 + \sin\theta$ at the point $\left(1 + \dfrac{\sqrt{3}}{2}, \dfrac{\pi}{3}\right)$?

(A) $\dfrac{1}{2}$

(B) $\dfrac{\sqrt{3}}{2}$

(C) $\dfrac{1 + \sqrt{3}}{1 - \sqrt{3}}$

(D) $1 + 2\sqrt{3}$

(E) $\dfrac{1 + 2\sqrt{3}}{1 - 2\sqrt{3}}$

BC 205. The velocity vector of an object moving in a plane is $\langle 3t^2, -9t \rangle$. What is the magnitude of the acceleration to the nearest tenth at time $= 3$?

(A) 10.0
(B) 18.2
(C) 20.1
(D) 27.3
(E) 54.2

206. A penny is dropped from the roof of a building 200 ft tall. The position function of the penny is $s(t) = -16t^2 + 200$, where $t \geq 0$ is in seconds. Approximating to the nearest second, find the time t when the penny will hit the ground.

(A) 2 s
(B) 4 s
(C) 7 s
(D) 10 s
(E) 16 s

207. What is the slope of the normal line to a curve at point (x, y) if the slope of the tangent at the same point is $\frac{1}{3}$?

(A) -3

(B) $-\frac{1}{3}$

(C) $-\frac{y}{3}$

(D) $\frac{y}{3}$

(E) 3

208. If f is a differentiable function and $f(5) = 10$ and $f'(5) = 2$, what is the approximate value of $f(5.5)$?

(A) 10

(B) 10.5

(C) 11

(D) 11.5

(E) 12

BC 209. A particle moving in a plane has a position defined by $\langle 2t^2 - 1, 4t \rangle$. What is the particle's speed at time $t = 1$?

(A) $2\sqrt{2}$

(B) $2\sqrt{5}$

(C) 4

(D) $4\sqrt{2}$

(E) $4\sqrt{5}$

210. The position function of a particle moving in a straight line is $s(t) = t^3 - 9t^2 + 24t - 2$. During what time interval is the particle moving to the left?

(A) $t \geq 0$

(B) $2 < t < 4$

(C) $2 \leq t \leq 4$

(D) $t > 4$

(E) $0 \leq t \leq 2$

211. What is the approximate value of $\cos 62°$ using linear approximation?

(A) 0.45

(B) 0.47

(C) 0.49

(D) 0.51

(E) 0.52

212. If the function y is differentiable at $x = a$ such that $\frac{dy}{dx}\big|_{x=a}$ does not exist but $\frac{dx}{dy}\big|_{x=a} = 0$, describe the tangents to function y at $x = a$.

(A) horizontal tangents
(B) parallel tangents
(C) vertical tangents
(D) no tangents
(E) horizontal and vertical tangents

213. Find the approximate value of $(5.2)^3$ using linear approximation.

(A) 130
(B) 135
(C) 140
(D) 145
(E) 150

214. A particle moving along a straight line has a velocity function of $v(t) = 2t^3 - \frac{1}{2}t^2 + 4t - 6$. What is its acceleration at time $t = 4$?

(A) 90
(B) 96
(C) 104
(D) 112
(E) 130

BC 215. What is the speed of a particle whose motion is defined by $y = -2t^2 + 5t$ and $x = 6t$, when $t = 1$?

(A) 7
(B) 9
(C) $\sqrt{37}$
(D) $\sqrt{57}$
(E) $\sqrt{117}$

BC 216. What is the slope of the tangent to the curve defined by $y = \frac{1}{3}t^2 - 7$ and $x = 6t - 1$, when $t = 3$?

(A) $\frac{1}{3}$
(B) $\frac{2}{3}$
(C) 1
(D) $\frac{3}{2}$
(E) 3

BC 217. The function $r = \langle t^3, 2t^2 \rangle$ defines an object moving in a plane. What is the vector T tangent to the path of the object at time $t = 2$?

(A) $\left\langle \dfrac{8}{\sqrt{3}}, \dfrac{32}{\sqrt{3}} \right\rangle$

(B) $\left\langle \dfrac{3}{\sqrt{13}}, \dfrac{2}{\sqrt{13}} \right\rangle$

(C) $\left\langle \dfrac{4}{\sqrt{13}}, \dfrac{4}{\sqrt{13}} \right\rangle$

(D) $\left\langle \dfrac{12}{\sqrt{13}}, \dfrac{8}{\sqrt{13}} \right\rangle$

(E) $\langle 4, 8 \rangle$

BC 218. A curve is defined by $x(t) = 2\sin t$ and $y(t) = t^2 - 2t$. Find $\dfrac{dy}{dx}$.

(A) $\dfrac{\sin t}{t - 1}$

(B) $\dfrac{t - 1}{\cos t}$

(C) $\dfrac{\cos t}{2(t - 1)}$

(D) $\dfrac{2t - 1}{\cos t}$

(E) $\dfrac{\sin t}{2t - 1}$

BC 219. The velocity of a particle $v = \langle 4t^2, 3t \rangle$. What is the magnitude of the acceleration to the nearest tenth at $= 2$?

(A) 3.7
(B) 8.5
(C) 11.1
(D) 16.3
(E) 17.1

220. Find the value of x at which the graphs of $y = x^2$ and $y = 4x$ have parallel tangents.

(A) $x = \dfrac{1}{2}$

(B) $x = 2$
(C) $x = 2\sqrt{2}$
(D) $x = 4$
(E) $x = 8$

BC 221. Calculate the slope of the line tangent to the curve $r = 1 + 2\cos\theta$ at the point $\left(2, \dfrac{\pi}{3}\right)$.

(A) $-2\sqrt{3}$

(B) $-\sqrt{3}$

(C) $-\dfrac{1}{\sqrt{3}}$

(D) $\dfrac{1}{2\sqrt{3}}$

(E) $\dfrac{1}{3\sqrt{3}}$

222. A ball is dropped from the roof of a building and hits the ground 15 s later. The position function of the ball is given as $s(t) = -16t^2 - v_0t + s_0$, where s_0 is the initial position measured in feet and v_0 is the initial velocity. Find the height of the building.

(A) 240 ft
(B) 900 ft
(C) 1225 ft
(D) 2400 ft
(E) 3600 ft

BC 223. A curve is defined by $x(t) = -3t^3 - 3t^2 + 6t - 1$ and $y(t) = 3\sin t$. Find $\dfrac{dy}{dx}$.

(A) $\dfrac{\sin t}{t^2 - t + 6}$

(B) $\dfrac{3\cos t}{2 - 2t - 3t^2}$

(C) $\dfrac{\cos t}{2 - 2t - 3t^2}$

(D) $-\dfrac{\sin t}{3t^2 - 2t + 2}$

(E) $-\dfrac{\cos t}{9t^2 + 6t - 6}$

224. The position function of a particle moving in a straight line is $s(t) = t^3 + t^2 - 5t + 1$, where $t \geq 0$. At what value(s) of t does the particle undergo a change of direction?

(A) $t = 1$

(B) $t = 1, \dfrac{3}{2}$

(C) $t = 1, \dfrac{3}{5}$

(D) $t = 1, \dfrac{5}{3}$

(E) $t = 1, 3$

225. If f is a differentiable function and $f(3) = \dfrac{1}{3}$ and $f'(3) = -\dfrac{1}{9}$, find the approximate value of $f(3, 4)$.

(A) 0.288

(B) 0.294

(C) 0.302

(D) 0.312

(E) 0.340

226. At what value(s) of x do the graphs of $y = 2x^3$ and $y = \dfrac{1}{2}x^2 + x + 6$ have parallel tangents?

(A) $x = 0, \dfrac{1}{3}, -\dfrac{1}{2}$

(B) $x = 0, \dfrac{1}{3}, \dfrac{1}{2}$

(C) $x = -\dfrac{1}{3}, \dfrac{1}{2}$

(D) $x = \dfrac{1}{3}, -\dfrac{1}{2}$

(E) $x = 2, -3$

227. Using linear approximation, find the approximate value of $\sqrt{139}$.

(A) 11.49

(B) 11.73

(C) 11.79

(D) 11.89

(E) 12.21

BC 228. Find the acceleration vector for a particle whose motion is defined by $x = -4t^2 + 2t - 1$ and $y = 6t$.

(A) $\langle 0, 6 \rangle$
(B) $\langle 8, 6 \rangle$
(C) $\langle -8, 6 \rangle$
(D) $\langle -8, 0 \rangle$
(E) $\langle 0, -8 \rangle$

BC 229. The motion of a particle is defined parametrically by $x(t) = t^3 - 5$ and $y(t) = \dfrac{1}{2}t^2 + 1$. Find the speed of the particle in terms of time t.

(A) $2t$

(B) $3t^2 + 1$

(C) $\sqrt{10t^4 + t}$

(D) $t\sqrt{9t^2 + 1}$

(E) $3t\sqrt{t^2 + 1}$

BC 230. What is the magnitude of the acceleration of an object moving in a plane whose position is defined by $r = \langle \sqrt{t}, t^3 \rangle$ at $t = 4$?

(A) 24
(B) 26

(C) $8\sqrt{5}$

(D) $2\sqrt{143}$

(E) $2\sqrt{145}$

231. What is the increase in acceleration of a particle whose velocity function is $v(t) = t^2 - 2t + 1$ over the interval $0 \le t \le 5$?

(A) 8
(B) 10
(C) 12
(D) 15
(E) 16

232. The line $y = x^2 - 2$ is tangent to the graph of $y = \dfrac{1}{2}x + a$. Find the value(s) of a.

(A) -1

(B) $-\dfrac{31}{16}$

(C) $-\dfrac{43}{16}$

(D) $-\dfrac{45}{16}$

(E) -3

BC 233. Write an equation for the line(s) tangent to the curve $\langle e^t, e^{-t} \rangle$ at the point $(1, 1)$.

(A) $y = 2 - x$
(B) $y = x - 1$
(C) $y = x - 2$
(D) $y = e^{-x}$
(E) $y = -e^x$

BC 234. Find $\dfrac{dy}{dx}$ if $x = r \tan \theta$ and $y = r \sec \theta$.

(A) $\dfrac{\tan \theta}{\sec \theta}$

(B) $\dfrac{r \sec \theta \tan \theta + \sec \theta \dfrac{dr}{d\theta}}{r \sec^2 \theta + \tan \theta \dfrac{dr}{d\theta}}$

(C) $\dfrac{dr}{d\theta} \left(\dfrac{\sec \theta}{\tan \theta} \right)$

(D) $\dfrac{dr}{d\theta} \left(\dfrac{\sec \theta \tan \theta + 1}{\sin^2 \theta + \tan \theta} \right)$

(E) $\dfrac{\sec \theta \left(r + \dfrac{dr}{d\theta} \right)}{(\sec \theta + \tan \theta) \left(\dfrac{dr}{d\theta} \right)}$

235. A ball is dropped from the roof of a house that is 40 ft high. The position of the ball can be described by the equation $s(t) = -16t^2 + 40$, where $t \geq 0$ is in seconds. When does the ball hit the ground?

(A) $\dfrac{\sqrt{5}}{4}$

(B) $\dfrac{\sqrt{10}}{2}$

(C) $\pm \dfrac{\sqrt{10}}{2}$

(D) $\sqrt{10}$

(E) $2\sqrt{10}$

236. Using linear approximation, what is the approximate value of $\sin 122°$?

(A) 0.826
(B) 0.849
(C) 0.859
(D) 0.866
(E) 0.883

237. The position function of a moving object is $s(t) = 3t^3 - t^2 + 1$. What is the instantaneous velocity of the object at time $t = 2$, where distance is measured in feet and time in seconds?

(A) 20 ft/s
(B) 21 ft/s
(C) 32 ft/s
(D) 36 ft/s
(E) 40 ft/s

238. Write an equation to the tangent line of $f(x) = x^3 + 2x - 1$ at $(-1, -4)$.

(A) $y = 5x + 1$
(B) $y = -5x - 2$
(C) $y = 5x - 8$
(D) $y = 5x - 4$
(E) $y = -5x - 1$

BC 239. A function is defined parametrically by $x(t) = 3\cos t$ and $y(t) = t^3 - 3t^2 + 1$. Find $\dfrac{dy}{dx}$.

(A) $-\dfrac{\sin t}{t^2 - 2}$

(B) $\dfrac{\sin t}{t^2 - t + 1}$

(C) $-\dfrac{t(t - 2)}{\sin t}$

(D) $-\dfrac{(t^2 - 2)}{\sin t}$

(E) $\dfrac{3t^2 - 6}{\sin t}$

240. A function is defined by $= x^2 - 1$. Find the equation of any horizontal tangent lines to the curve.

(A) $y = -2$
(B) $y = -1$
(C) $y = 1$
(D) $y = 2$
(E) $y = 0,\ 2$

241. What is the speed of a particle whose motion is defined by $x(t) = 2t^2 - 3$ and $y(t) = -5t$ when $t = 2$?

(A) $5\sqrt{2}$

(B) $2\sqrt{10}$

(C) $\sqrt{39}$

(D) 7

(E) $\sqrt{89}$

242. What are the velocity and acceleration of a particle with position function $s(t) = 5t^2 - 12t + 1$ at $t = 2$ if distance is measured in feet and time in seconds?

    (A) $v = 8$ ft/s, $a = -10$ ft/s²
    (B) $v = 8$ ft/s, $a = 10$ ft/s²
    (C) $v = 10$ ft/s, $a = 10$ ft/s²
    (D) $v = 12$ ft/s, $a = 10$ ft/s²
    (E) $v = 20$ ft/s, $a = 12$ ft/s²

243. Find the slope of the normal line to the function $y = 3\cos 2x$ at the point $x = \dfrac{\pi}{4}$.

    (A) $-\dfrac{1}{6}$

    (B) $\dfrac{1}{6}$

    (C) $\dfrac{1}{3\sqrt{2}}$

    (D) $-3\sqrt{2}$

    (E) -6

244. f is a differentiable function and $f(4) = 48$ and $f'(4) = 48$. Using tangent line approximation, what is the approximate value of $f(4.3)$?

    (A) 49
    (B) 50.2
    (C) 55.2
    (D) 55.5
    (E) 151.2

245. The position function of an object in horizontal motion is $s(t) = 3t^3 - 3t^2 - 1$. At what time(s) t does its acceleration $= 0$?

    (A) $t = 0$

    (B) $t = \dfrac{1}{3}$

    (C) $t = \pm\dfrac{1}{3}$

    (D) $t = 0, 3$

    (E) $t = 2, 3$

246. A curve is defined by the equation $y = x^3 - 4$.

    (A) Find the slope of the tangent to the curve at point $(2, 4)$.
    (B) Find the slope of the normal line at point $(2, 4)$.
    (C) Write an equation for the normal line.

247. A projectile is fired straight upward with a velocity of 256 ft/s. Its distance from the ground after being fired is given by $s(t) = -16t^2 + 256t$, where t is the time in seconds since the projectile was fired.

    (A) Write a velocity function for the projectile.
    (B) What is the maximum altitude reached by the projectile?
    (C) What is the acceleration at any time t?
    (D) At what time does the projectile hit the ground?

BC 248. An object is moving on a path defined by the vector-valued function $r(t) = \langle 2\cos t, \ 2\sin t \rangle$. Find the following functions at time $t = \dfrac{\pi}{4}$.

    (A) the velocity vector
    (B) the acceleration vector
    (C) the tangent vector
    (D) the normal vector

BC 249. A curve is defined by parametric equations $x(t) = t^3 - 2t + 1$ and $y(t) = t^2 - 5$.

    (A) Find an equation for the tangent line corresponding to $t = 1$.
    (B) For what value(s) of t is the tangent line horizontal?
    (C) For what value(s) of t is the tangent line vertical?

250. A penny is dropped from a building 150 ft tall. The position function of the penny is $s(t) = -16t^2 + 150$, where $t \geq 0$ is in seconds. Find the following:

    (A) the instantaneous velocity of the penny at $t = 1$ s
    (B) the average velocity for the first 2 s
    (C) the time when the penny will hit the ground

CHAPTER 6

Integration

251. Evaluate the integral $\int (x^4 - 3x^2 + 1)\,dx$.

(A) $\dfrac{x^5}{5} - \dfrac{3x^3}{2} + x + C$

(B) $4x^3 - 6x + C$

(C) $\dfrac{x^4}{4} - \dfrac{3x^2}{2} + x + C$

(D) $\dfrac{x^5}{5} - x^3 + x + C$

(E) $\dfrac{x^5}{5} - \dfrac{x^3}{3} - x + C$

BC 252. Evaluate $\displaystyle\int \dfrac{1}{x(x+2)}\,dx$.

(A) $\ln|x(x+2)| + C$

(B) $\dfrac{1}{2}\ln|x| - \dfrac{1}{2}\ln|x+2| + C$

(C) $\dfrac{1}{2x^2(x+2)^2} + C$

(D) $\dfrac{2}{x(x+2)^2} + C$

(E) $\dfrac{1}{2}\ln|x(x+2)| + C$

253. Evaluate $\int \dfrac{1}{2} x^2 \sin x^3 \, dx$.

(A) $-\dfrac{1}{6} \cos x^3 + C$

(B) $-\dfrac{1}{6} x^3 \cos x^3 + C$

(C) $\dfrac{1}{2} x^4 \sin x + C$

(D) $\dfrac{3}{2} x^2 \sin x^3 + C$

(E) $\dfrac{1}{6} \cos x^3 + C$

254. Evaluate $\int \dfrac{1}{x^2 + 6x + 13} \, dx$.

(A) $\ln \left| (x+3)^2 + 4 \right| + C$

(B) $-\dfrac{1}{2} \tan\left(\dfrac{x+3}{2} \right) + C$

(C) $-\left(\dfrac{1}{3x^3} + \dfrac{1}{12x^2} + \dfrac{1}{13x} \right) + C$

(D) $-\left(\dfrac{1}{x^2 + 6x + 13} \right)^{-2} + C$

(E) $\dfrac{1}{2} \tan\left(\dfrac{x+3}{2} \right) + C$

255. Evaluate $\int \dfrac{x^2 + 3x - 10}{x - 2} \, dx$.

(A) $\ln |x - 2| \left(\dfrac{1}{3} x^3 + \dfrac{3}{2} x^2 - 10x \right) + C$

(B) $\dfrac{\dfrac{1}{3} x^3 + \dfrac{3}{2} x^2 - 10x}{(x-2)^2} + C$

(C) $\dfrac{1}{2} x^2 + 5x + C$

(D) $(2x + 3) \ln |x - 2| + C$

(E) $\left(\dfrac{1}{3} x^3 + \dfrac{3}{2} x^2 - 10x \right) \left(\dfrac{1}{x - 2} \right) + C$

256. Evaluate $\int 2xe^{(x^2+1)}dx$.

(A) $x^2e^{(2x+1)}+C$

(B) $e^{(x^2+1)}+C$

(C) $4x^2e^{(2x+1)}+C$

(D) $2e^{2x}+C$

(E) $2xe^{(2x+1)}+C$

257. Evaluate $\int 3x\cos 3x\,dx$.

(A) $x\cos 3x-\dfrac{1}{3}\sin 3x+C$

(B) $\dfrac{1}{3}x\cos 3x-3x\sin 3x+C$

(C) $\dfrac{1}{9}x\sin 3x+C$

(D) $x\sin 3x+\dfrac{1}{3}\cos 3x+C$

(E) $3\sin 3x+C$

BC 258. Evaluate $\displaystyle\int \frac{dx}{x^2-5x-6}$.

(A) $\dfrac{1}{7}\ln|x-6|-\dfrac{1}{7}\ln|x+1|+C$

(B) $\ln|x^2-5x-6|+C$

(C) $\dfrac{1}{3x^3}-\dfrac{1}{10x^2}-\dfrac{1}{6x}+C$

(D) $\dfrac{1}{6}\ln|x-6|-\dfrac{1}{6}\ln|x+1|+C$

(E) $\dfrac{1}{7}\ln|x+1|-\dfrac{1}{7}\ln|x+6|+C$

BC 259. Evaluate $\int e^x\sin x\,dx$.

(A) $-e^x\cos x+C$

(B) $e^x(\sin x+\cos x)+C$

(C) $\dfrac{1}{2}e^x(\cos x-\sin x)+C$

(D) $2e^x\sin x+C$

(E) $\dfrac{1}{2}e^x(\sin x-\cos x)+C$

260. Evaluate $\int \dfrac{5x^2 - 2x + 1}{x^2} dx$.

(A) $5x - \dfrac{1}{x^2} + \dfrac{1}{3x^3} + C$

(B) $\dfrac{\dfrac{5}{3}x^3 - x^2 + x}{x^3} + C$

(C) $5x - 2\ln|x| + \dfrac{1}{x} + C$

(D) $-\dfrac{1}{3}x^3\left(\dfrac{5}{3}x^3 - x^2 + x\right) + C$

(E) $5x - 2\ln|x| - \dfrac{1}{x^2} + C$

261. Evaluate $\int(5\sec x)(2\tan x)dx$.

(A) $10\sec x + C$
(B) $10\sec^{-2} x + C$
(C) $-10\sec x + C$
(D) $10\tan^2 x + C$
(E) $-10\tan x \sec x + C$

262. Evaluate $\int 5^x dx$.

(A) $5^x + C$

(B) $\dfrac{5^x}{\ln(5)} + C$

(C) $\ln(5)5^x + C$

(D) $\dfrac{5^{x+1}}{x+1} + C$

(E) $5^x + \ln 5 + C$

263. Evaluate $4\int(\ln x + \sin^2 x)dx$

(A) $4x\ln|x| - \sin 2x + C$
(B) $4(x\ln|x| - x - \cos 2x) + C$
(C) $4x\ln|x| - 2x - \sin 2x + C$
(D) $4(x\ln|x| + \cos 2x) + C$
(E) $4x(\ln|x| - \sin 2x - 1) + C$

264. Evaluate $\displaystyle\int \frac{x^2}{(x^3+1)^3}\,dx$.

    (A) $\displaystyle\frac{x^3}{12(x^3+1)^4}+C$

    (B) $\displaystyle\frac{x^3}{6(x^3+1)^2}+C$

    (C) $-\displaystyle\frac{1}{12(x^3+1)^4}+C$

    (D) $-\displaystyle\frac{1}{6(x^3+1)^2}+C$

    (E) $\displaystyle\frac{1}{9(x^3+3)^3}+C$

265. Evaluate $\displaystyle\int \frac{\sin x - \cos x}{\cos x}\,dx$.

    (A) $\ln|\sec x| - x + C$

    (B) $-\displaystyle\frac{(\cos x + \sin x)}{\cos x}+C$

    (C) $-\displaystyle\frac{(\cos x - \sin x)}{\sin x}+C$

    (D) $\ln|\sec x| + x + 1 + C$

    (E) $\tan x - x + C$

BC 266. Evaluate $\displaystyle\int 3x\sec^2 x\,dx$.

    (A) $\displaystyle\frac{3}{2}x^2\tan x + C$

    (B) $3x\tan x + C$

    (C) $3x\ln|\sec x| - 3\tan x + C$

    (D) $3x\tan x - 3\ln|\sec x| + C$

    (E) $\displaystyle\frac{3}{2}x^2\tan x - 3\ln|\sec x| + C$

267. Evaluate $\displaystyle\int \frac{2x+3}{x}\,dx$.

    (A) $-\displaystyle\frac{(x^2+3x)+C}{x^2}$

    (B) $-(x^2+3x)\ln|x| + C$

    (C) $3\ln|x| - 2x + C$

    (D) $x^2+3x+3\ln|x| + C$

    (E) $2x+3\ln|x| + C$

268. Evaluate $\int 7^{3x} \, dx$.

(A) $\dfrac{7^{3x}}{3 \ln 7} + C$

(B) $\dfrac{1}{3} 7^{3x} + C$

(C) $\dfrac{7^{3x}}{\ln 7} + C$

(D) $(3x+1)7^{3x+1} + C$

(E) $\dfrac{7^{3x+1}}{3x+1} + C$

BC 269. Evaluate $\displaystyle \int \dfrac{dx}{x^2 - 4}$.

(A) $\dfrac{1}{4} \ln|x-2| + \dfrac{1}{4} \ln|x+2| + C$

(B) $\ln|x^2 - 4| + C$

(C) $\dfrac{1}{4} \ln|x-2| - \dfrac{1}{4} \ln|x+2| + C$

(D) $-\dfrac{1}{2}(x^2 - 4)^{-2} + C$

(E) $x^2 - 4 + C$

270. Evaluate $\dfrac{1}{2} \displaystyle \int 4^x \ln(4) \, dx$.

(A) $\dfrac{4^x}{2 \ln(4)} + C$

(B) $\dfrac{1}{2}\left[4^x \ln(4) - \dfrac{4^{x+1} \ln 4}{x+1} \right] + C$

(C) $4^x \ln(4) + C$

(D) $\dfrac{1}{2}[4^x \ln(4) - 4^x] + C$

(E) $\dfrac{1}{2} 4^x + C$

271. Evaluate $\int (x^2 - \sin^2 x)\, dx$.

(A) $\dfrac{1}{3}x^3 - \dfrac{1}{3}\sin^3 x + C$

(B) $\dfrac{1}{3}x^3(1 - \sin 2x) + C$

(C) $\dfrac{1}{3}x^3(1 - \cos 2x) + C$

(D) $\dfrac{1}{3}x^3 - \dfrac{1}{2}x + \dfrac{1}{4}\sin 2x + C$

(E) $\dfrac{1}{3}x^3 - \dfrac{1}{2}x - \dfrac{1}{4}\cos 2x + C$

272. Evaluate $\displaystyle\int \dfrac{4}{x\sqrt{x^2 - 4}}\, dx$.

(A) $2\sin^{-1}\left|\dfrac{x}{2}\right| + C$

(B) $\dfrac{1}{2}\sec^{-1}\left|\dfrac{x}{2}\right| + C$

(C) $2\sec^{-1}\left|\dfrac{x}{2}\right| + C$

(D) $4\ln|x| + 2\sec^{-1}\left|\dfrac{x}{2}\right| + C$

(E) $4\ln|x| + 2\sin^{-1}\left|\dfrac{x}{2}\right| + C$

273. Evaluate $\int x(x-3)^3\, d$.

(A) $\dfrac{1}{5}(x-3)^5 + \dfrac{3}{4}(x-3)^4 + C$

(B) $\dfrac{1}{4}x(x-3)^4 + C$

(C) $\dfrac{1}{4}(x-3)^4 - (x-3)^3 + C$

(D) $\dfrac{1}{8}x^2(x-3)^4 + C$

(E) $\dfrac{1}{4}(x-3)^4 + \dfrac{1}{3}(x-3)^3 + C$

274. Evaluate $\int 3x^2 e^{(x^3-1)} dx$.

(A) $x^3 e^{(x^3-1)} + C$

(B) $e^{(x^3-1)} + C$

(C) $9x^2 e^{(x^3-1)} + C$

(D) $x^2 e^{3x^2} + C$

(E) $x^3 e^{3x^2} + C$

275. Evaluate $\int \dfrac{x^2}{\sqrt{1-x^6}} d$.

(A) $\dfrac{1}{3}\sin^{-1}(x^3) + C$

(B) $\dfrac{1}{3}x^3 \sin^{-1}(x^3) + C$

(C) $3x \sec^{-1}(x^3) + C$

(D) $\dfrac{1}{3}x \sec^{-1}(x^3) + C$

BC 276. Evaluate $\int x^2 e^{-x} dx$.

(A) $e^{-x}(2 + 2x - x^2) + C$

(B) $\dfrac{1}{3}x^3 e^{-x} + C$

(C) $-e^{-x}(x^2 + 2x + 2) + C$

(D) $-\dfrac{1}{3}x^3 e^{-x} + C$

(E) $e^{-x}(x^2 - 2x - 2) + C$

277. Evaluate $\int 9^{3x} dx$.

(A) $\dfrac{1}{3}9^{3x} + C$

(B) $(3x)9^{3x} + C$

(C) $\dfrac{1}{3x}(9^{3x}) + C$

(D) $(3)\dfrac{9^{3x}}{\ln 9} + C$

(E) $\left(\dfrac{1}{3}\right)\dfrac{9^{3x}}{\ln 9} + C$

278. Evaluate $\int \sqrt{x}(x^3 + 1)d$.

(A) $\dfrac{5}{2}x^{\frac{5}{2}} + \dfrac{3}{2}x^{\frac{3}{2}} + C$

(B) $\dfrac{7}{2}x^{\frac{7}{2}} + \dfrac{3}{2}x^{\frac{3}{2}} + C$

(C) $\dfrac{3}{2}x^{\frac{3}{2}}\left(\dfrac{1}{4}x^4 + 1\right) + C$

(D) $\dfrac{\sqrt{x}(2x^4 + 6x)}{9} + C$

(E) $\dfrac{1}{2}(\sqrt{x} + 3) + C$

279. Evaluate $\int \sec^3 x\, dx$.

(A) $\ln |\sec x \tan x|^2 + C$

(B) $\dfrac{1}{4}\ln |\sec x + \tan x|^4 + C$

(C) $\dfrac{1}{2}(\sec x \tan x + \ln |\sec x + \tan x|) + C$

(D) $\dfrac{1}{2}\ln |\sec x + \tan x| + C$

(E) $\dfrac{1}{2}(\sec x \tan x) + C$

280. Evaluate $\int \left(x^4 + \dfrac{x}{9 + x^4} \right) dx$.

(A) $\dfrac{1}{5}x^5 + \tan^{-1} x + C$

(B) $\dfrac{1}{5}x^5 + \dfrac{1}{6}\tan^{-1}\left(\dfrac{x^2}{3}\right) + C$

(C) $\dfrac{1}{5}x^5 + \sin\left(\dfrac{x^2}{3}\right) + C$

(D) $\dfrac{1}{5}x^5 + \dfrac{1}{3}\tan^{-1}\left(\dfrac{x^2}{3}\right) + C$

(E) $\dfrac{1}{5}x^5 + \dfrac{1}{9}\tan^{-1}\left(\dfrac{x}{9}\right) + C$

281. Evaluate $\int x(x-5)(x+2)dx$.

(A) $\dfrac{1}{3}x^3 - \dfrac{3}{2}x^2 - 10x + C$

(B) $\dfrac{1}{4}x^4 - 3x^3 - 10x^2 + C$

(C) $\dfrac{1}{4}x^4 - x^3 - 5x^2 + C$

(D) $-\dfrac{1}{2}x^2(x^2 + 3x + 10) + C$

(E) $\dfrac{1}{2}x^2(x^2 - 3x - 10) + C$

282. Evaluate $\displaystyle\int \dfrac{\sin\sqrt{x}}{\sqrt{x}}dx$.

(A) $-2\cos\sqrt{x} + C$

(B) $-\dfrac{1}{2}\cos\sqrt{x} + C$

(C) $-\dfrac{1}{2}\dfrac{\cos\sqrt{x}}{\sqrt{x}} + C$

(D) $-\dfrac{3}{2}\dfrac{\cos\sqrt{x}}{\sqrt{x^3}} + C$

(E) $-2\dfrac{\cos\sqrt{x}}{\sqrt{x^3}} + C$

283. Evaluate $\displaystyle\int \dfrac{3x^2 + x - 6}{x^2}dx$.

(A) $\ln(x^2)\left(x^3 + \dfrac{1}{2}x^2 - 6x\right) + C$

(B) $\dfrac{x^3 + \dfrac{1}{2}x^2 - 6}{x^3} + C$

(C) $3x + \ln x - \dfrac{6}{x} + C$

(D) $3x + \ln x + \dfrac{6}{x} + C$

(E) $3x + \dfrac{6}{x} + C$

284. Evaluate $\int 2x\sqrt{1+x^2}\,dx$.

(A) $-\dfrac{2}{\sqrt{1+x^2}}+C$

(B) $-\dfrac{2}{3\sqrt{(1+x^2)^3}}+C$

(C) $-\dfrac{2}{x^2\sqrt{1+x^2}}+C$

(D) $x^2\sqrt{(1+x^2)^3}+C$

(E) $\dfrac{2}{3}\sqrt{(1+x^2)^3}+C$

BC 285. Evaluate $\displaystyle\int \dfrac{3x+5}{x^2+3x-4}\,dx$.

(A) $-\dfrac{5}{4}x+C$

(B) $\dfrac{\dfrac{1}{3}x^3+\dfrac{3}{2}x^2+5x}{\dfrac{1}{3}x^3+\dfrac{3}{2}x^2-4x}+C$

(C) $\dfrac{8}{5}\ln|x+4|-\dfrac{7}{5}\ln|x-1|+C$

(D) $\dfrac{7}{5}\ln|x+4|-\dfrac{8}{5}\ln|x-1|+C$

(E) $\left(\dfrac{1}{3}x^3+\dfrac{3}{2}x^2+5x\right)\ln|x^2-3x-4|+C$

286. Evaluate $\int e^{(5x-1)}\,dx$.

(A) $(5x-1)e^{(5x-1)}+C$

(B) $\dfrac{1}{5}e^{(5x-1)}+C$

(C) $\dfrac{1}{5}e^{5x}+C$

(D) $\dfrac{e^{(5x-1)}}{5x-1}+C$

(E) $\dfrac{e^{5x}}{5x}+C$

287. Evaluate $\int 21^{3x}\, dx$.

(A) $\dfrac{1}{3}\ln(21)21^{3x}+C$

(B) $(3x)21^{3x}+C$

(C) $\dfrac{21^{3x}}{3\ln(21)}+C$

(D) $\dfrac{21^{3x}}{3x}+C$

(E) $\dfrac{1}{3}21^{3x}+C$

288. Evaluate $\int x(x-1)^6\, dx$.

(A) $\dfrac{1}{7}x(x-1)^7+C$

(B) $\dfrac{1}{14}x(x-1)^7+C$

(C) $\dfrac{1}{14}x^2(x-1)^7+C$

(D) $\dfrac{1}{8}(x-1)^8+\dfrac{1}{7}(x-1)^7+C$

(E) $\dfrac{1}{7}(x-1)^7+\dfrac{1}{6}(x-1)^6+C$

289. Evaluate $\int e^x(1-e^{2x})dx$.

(A) $e^x-\dfrac{1}{3}e^{3x}+C$

(B) $(e^{x-1})(1-e^{2x})+C$

(C) e^x-1+C

(D) $\dfrac{1}{3}e^x(1-e^{2x})+C$

(E) $e^x-e^{3x}+C$

290. Evaluate $\int \ln(e^{(x^2-x+1)})dx$.

(A) $\ln\left|e^{(x^2-x+1)}\right|+C$

(B) $\dfrac{1}{3}x^3-\dfrac{1}{2}x^2+x+C$

(C) $x^2-x+1+C$

(D) $e^{(x^2-x+1)}[\ln\left|e^{x^2-x+1}\right|-1]+C$

(E) $e^{(x^2-x+1)}+C$

291. Evaluate $\displaystyle\int \frac{x^2}{x^3+1}\,dx$.

(A) $\dfrac{1}{3}x^3 \ln|x^3+1| + C$

(B) $x^2 \ln|x^3+1| + C$

(C) $\dfrac{1}{3}\ln|x^3+1| + C$

(D) $\dfrac{1}{3}x^3 \dfrac{1}{(x^3+1)^2} + C$

(E) $\dfrac{1}{3}x^3(x^3+1) + C$

BC 292. Evaluate $\int \sin^4 x\,dx$.

(A) $\dfrac{1}{4}\left(x - \dfrac{1}{2}\sin(2x)\right) + C$

(B) $-\cos x \sin^3 x - \dfrac{1}{2}\sin(2x) + C$

(C) $-\cos x \sin^3 x + \dfrac{3x}{2} - \dfrac{3\sin(2x)}{4} + C$

(D) $-\dfrac{1}{4}\cos x \sin^3 x + \dfrac{3x}{8} - \dfrac{3}{8}\sin x \cos x + C$

(E) $-\dfrac{1}{4}\left(\cos x \sin^3 x + \dfrac{3x}{2} + 3\sin x \cos x\right) + C$

293. Evaluate $\displaystyle\int \frac{\ln x}{4x}\,dx$.

(A) $\dfrac{1}{8}(\ln x)^2 + C$

(B) $\dfrac{1}{4}\ln(x^2) + C$

(C) $\dfrac{1}{4}(\ln x)^2 + C$

(D) $\dfrac{1}{4}x \ln|x|\,(\ln|x| - 1) + C$

(E) $\dfrac{1}{4}\ln|x|\,(\ln|x| - 1) + C$

294. Evaluate $\displaystyle\int \frac{1}{x^2 - 6x + 13} dx$.

(A) $\ln |x^2 - 6x + 13| + C$

(B) $\dfrac{1}{\frac{1}{3}x^3 - 3x^2 + 13x} + C$

(C) $\dfrac{1}{3}\sin^{-1}\left(\dfrac{x-3}{2}\right) + C$

(D) $\dfrac{1}{3}\tan^{-1}\left(\dfrac{x-3}{3}\right) + C$

(E) $\dfrac{1}{2}\tan^{-1}\left(\dfrac{x-3}{2}\right) + C$

295. Evaluate $\displaystyle\int \csc^2 x \cot x \, dx$.

(A) $-\dfrac{1}{2}\cot x + C$

(B) $-2\csc x + C$

(C) $-\dfrac{1}{2}\cot^2 x + C$

(D) $-2\csc^2 x + C$

(E) $-\dfrac{1}{2}\csc^2 x \cot x + C$

BC 296. The slope of a function $f(x)$ at any point (x, y) is $\dfrac{x-3}{x^2 - 3x - 4}$.

The point $\left(5, \dfrac{4}{5}\ln 6\right)$ is on the graph of $f(x)$.

(A) Write an equation of the tangent line to the graph of $f(x)$ at $x = 5$.

(B) Use the tangent line in part A to approximate $f(4.5)$ to the nearest thousandth.

(C) Find the antiderivative of $\dfrac{df}{dx} = \dfrac{x-3}{x^2 - 3x - 4}$ with the condition $f(5) = \dfrac{4}{5}\ln 6$.

(D) Use the result of part C to find $f(4.5)$ to the nearest thousandth.

297. A particle moves such that its acceleration $a(t) = 1 + \dfrac{1}{\sqrt{t}}$. The initial conditions are $v(0) = 2$ and $s(0) = 10$.

(A) Find an expression for $v(t)$.
(B) Find the position function $s(t)$.

298. A bacteria population is growing at a rate of $200 + 9\sqrt{t} - 5t$, where t is the time given in hours.

(A) The population size, $P(t)$, is the antiderivative of the rate of growth. Find $P(t)$.
(B) If it is given that the initial population, $P(0) = 2000$, find the constant C from the integration in part A.
(C) Find the population, $P(t)$, after $t = 4$ h.

299. The marginal revenue that a manufacturer receives for goods, q, is given by $MR = 100 - 0.5q$.

(A) Find the antiderivative of MR to get a function for the total revenue.
(B) How many goods must be produced to generate a total revenue of $500?
(C) At what point in production will you reach the point of diminishing return, when revenue begins to decrease?

300. The density ρ of a 6 m long metal rod of nonuniform density is given by $\rho(x) = \dfrac{3}{2}\sqrt{x}$ in units of kg/m and x is given as the distance along the rod measuring from the left end ($x = 0$).

(A) Find $m(x)$, the mass, as the antiderivative of $\rho(x)$.
(B) What is the mass 1 m from the left end to the nearest tenth of a kilogram?
(C) What is the total mass of the rod to the nearest tenth of a kilogram?

Definite Integrals

301. Evaluate $\sum_{i=1}^{5} i^3$.

  (A) 99
  (B) 124
  (C) 125
  (D) 225
  (E) 325

302. Evaluate $\int_{0}^{4} (x^3 - 3x + 1)\, dx$.

  (A) −4
  (B) 0
  (C) 24
  (D) 32
  (E) 44

303. Evaluate $\int_{\frac{\pi}{2}}^{\pi} \frac{1}{4 + x^2}\, dx$.

  (A) 0.169
  (B) 0.334
  (C) 0.338
  (D) 0.535
  (E) 0.835

BC 304. Evaluate $\int_{0}^{4} \frac{1}{\sqrt{16 - x^2}}\, dx$.

  (A) 0
  (B) $\dfrac{\pi}{4}$
  (C) $\dfrac{\pi}{2}$
  (D) 2
  (E) π

305. Evaluate $\displaystyle\int_0^{2\pi} (x - \cos x)\,dx$.

(A) 0

(B) $2\pi^2 - \pi$

(C) $2\pi^2 - \dfrac{\pi}{2}$

(D) $2\pi^2$

(E) $4\pi^2$

306. If $\displaystyle\int_0^k (5 - x)\,dx = -12$ and $k > 0$, find k.

(A) -2

(B) 0

(C) 2

(D) 12

(E) 15

307. Evaluate $\displaystyle\sum_{i=1}^{n} n^2 i(i + 2)$.

(A) $n^3(n+1)\left(\dfrac{1}{3}n + \dfrac{7}{6}\right)$

(B) $n^3(n+1)(2n+1)$

(C) $\dfrac{1}{2}n^2(n+1)^2(2n+1)$

(D) $\dfrac{1}{6}n^3(n+1)(2n+1)$

(E) $n^3(n+1)(3n+2)$

308. Evaluate $\displaystyle\int_0^{\frac{\pi}{2}} \sin(2x)\,dx$.

(A) -1

(B) 0

(C) 1

(D) $\dfrac{\pi}{2}$

(E) π

309. Find $\frac{dy}{dx}$ if $y = \int_x^{x^2} (t^2 - t + 1)dt$.

(A) $2x^5 - 2x^3 - \frac{1}{2}x^2 - x + 1$

(B) $2x^5 - 2x^3 - x^2 + 3x - 1$

(C) $2x^5 - 2x^3 + x^2 - 3x + 1$

(D) $\frac{1}{3}x^6 - \frac{1}{2}x^4 - \frac{1}{3}x^3 + \frac{3}{2}x^2 - x$

(E) $\frac{1}{3}x^6 - \frac{1}{2}x^4 - \frac{1}{3}x^3 + \frac{3}{2}x^2 - x$

310. Evaluate $\int_0^6 |2x - 4|\, dx$.

(A) 12
(B) 20
(C) 24
(D) 36
(E) 60

311. Evaluate $\int_0^{\frac{\pi}{2}} \frac{\sin x}{\cos x + 1}\, dx$.

(A) 0
(B) −0.693
(C) 0.693
(D) 0.500
(E) 1.693

BC 312. Evaluate $\int_{-\infty}^0 e^x dx$.

(A) −1
(B) 0
(C) 1
(D) e
(E) Integral is divergent.

313. Find k if $\int_{-4}^k (2x - 3)dx = -30$ and $k > 1$.

(A) 1
(B) 2
(C) 3
(D) 4
(E) 5

314. Find $\dfrac{dy}{dx}$ if $y = \displaystyle\int_x^{x^2} \dfrac{1}{\sqrt{t}}\, dt$.

(A) $2\sqrt{2}(x-1)$

(B) $2\sqrt{2}(x-\sqrt{x})$

(C) $2 - \dfrac{1}{\sqrt{x}}$

(D) $\dfrac{2\sqrt{2}}{x} - \dfrac{1}{\sqrt{x}}$

(E) $2\sqrt{2}(\sqrt{x}-1)$

315. Evaluate $\displaystyle\int_{-2}^{2} (x^2 - 2x + 1)\, dx$.

(A) $1\dfrac{1}{3}$

(B) $2\dfrac{2}{3}$

(C) $4\dfrac{1}{3}$

(D) $6\dfrac{2}{3}$

(E) $9\dfrac{1}{3}$

316. Evaluate the sum $\displaystyle\sum_{i=0}^{6} i^2(i-1)$.

(A) 240

(B) 280

(C) 328

(D) 350

(E) 455

317. Evaluate $\displaystyle\int_{0}^{\frac{\pi}{2}} \sin^2 x\, dx$.

(A) $-\dfrac{\pi}{4}$

(B) 0

(C) $\dfrac{\pi}{4}$

(D) $\dfrac{\pi}{2}$

(E) $\dfrac{3\pi}{2}$

318. Evaluate $\int_0^1 x(x+1)^{\frac{1}{3}} dx$.

(A) −0.591

(B) −0.321

(C) 0.231

(D) 0.321

(E) 0.591

319. Evaluate $\int_{-6}^6 |x^2 - 9| dx$.

(A) 27

(B) 54

(C) 90

(D) 108

(E) 180

320. Evaluate $\int_1^4 \frac{e^{\sqrt{x}}}{\sqrt{x}} dx$.

(A) $2e(e-1)$

(B) $2(e^2 - 1)$

(C) $2(e^2 + 1)$

(D) $2e(e+1)$

(E) $2e^2$

321. Evaluate $\int_{\ln 2}^{\ln 3} xe^{x^2} dx$.

(A) 0

(B) 0.5

(C) 0.707

(D) 0.864

(E) 1.672

322. Evaluate $\int_0^3 |4x - 2| dx$.

(A) 9

(B) 12

(C) 13

(D) 15

(E) 18

323. Evaluate $\displaystyle\int_0^{\frac{\pi}{3}} \tan^2 x \sec^2 x \, dx$.

(A) 0

(B) $\dfrac{\sqrt{3}}{3}$

(C) $\sqrt{3}$

(D) $\dfrac{\pi^2}{9}$

(E) $\dfrac{\pi^3}{27}$

324. Find k if $\displaystyle\int_0^k \left(2x - \dfrac{\sqrt{x}}{2} \right) dx = 0$ and $k > 0$.

(A) $\dfrac{1}{9}$

(B) $\dfrac{1}{3}$

(C) $\dfrac{4}{9}$

(D) $\dfrac{2}{3}$

(E) 1

BC 325. Evaluate $\displaystyle\int_2^3 \dfrac{1}{\sqrt{x-2}} \, dx$.

(A) 0

(B) $\dfrac{1}{2}$

(C) 1

(D) 2

(E) ∞

326. Evaluate $\displaystyle\int_0^1 10^{3x} \, dx$.

(A) 136.90

(B) 144.62

(C) 144.90

(D) 152.22

(E) 160.80

327. Evaluate $\int_e^{e^2} \frac{1}{x \ln x} dx$.

(A) 0

(B) e

(C) 1

(D) $e^2 - e$

(E) ln 2

328. Find $\frac{dy}{dx}$ if $y = \int_0^{x^2} \frac{1}{2} \cos t \ dt$.

(A) $x \cos x^2$

(B) $x \sin x^2$

(C) $2x \cos x^2$

(D) $2x \sin x^2$

(E) $x \cos (2x)$

329. Evaluate $\int_4^9 \frac{x+1}{\sqrt{x}} dx$.

(A) $9\frac{1}{3}$

(B) $13\frac{1}{3}$

(C) $14\frac{2}{3}$

(D) $15\frac{1}{6}$

(E) $33\frac{1}{3}$

330. Evaluate $\int_{-\infty}^{\infty} \frac{1}{1+x^2} dx$.

(A) $-\infty$

(B) $\frac{\pi}{2}$

(C) 0

(D) π

(E) ∞

331. If $f(x) = \int_{\frac{\pi}{2}}^{x} \tan^{-1} t \, dt$, find $f'\left(\frac{\pi}{6}\right)$ to the nearest thousandth.

(A) -0.4823

(B) 0.4823

(C) 0.5236

(D) 0.5774

(E) 1.486

332. Evaluate $\int_{-1}^{1} (x^2 - 3)(x^5 + 2) dx$.

(A) $-13\frac{1}{3}$

(B) -12

(C) $-10\frac{2}{3}$

(D) $1\frac{1}{3}$

(E) $10\frac{2}{3}$

333. Evaluate $\int_{0}^{\infty} \frac{1}{x^2 - 3x + 2} dx$.

(A) $-\infty$

(B) $-\ln 2$

(C) 0

(D) $\ln 2$

(E) ∞

334. Evaluate $\int_{0}^{1} \frac{x^2}{\sqrt{1 - x^6}} dx$.

(A) $-\frac{\pi}{6}$

(B) 0

(C) $\frac{\pi}{6}$

(D) $\frac{\pi}{3}$

(E) $\frac{\pi}{2}$

335. Find $\dfrac{dy}{dx}$ if $y = \displaystyle\int_{\sin x}^{\cos x}\left(1 - \dfrac{1}{2}t\right)dt$.

(A) $\sin x \cos x - \sin x - \cos x$

(B) $2\sin x \cos x$

(C) $2 - \sin x \cos x$

(D) $\sin x \cos x$

(E) $\cos x - \sin x - \sin x \cos x$

336. Evaluate $\displaystyle\int_{\frac{\pi}{6}}^{\frac{\pi}{4}} \csc^2 x \cot x \, dx$.

(A) $\dfrac{1}{2} - \sqrt{3}$

(B) $-\dfrac{\sqrt{3}}{2}$

(C) 0

(D) 1

(E) $-\dfrac{1}{2} + \sqrt{3}$

337. Evaluate $\displaystyle\int_{1}^{e} \dfrac{\ln x}{5x} \, dx$.

(A) $\dfrac{1}{10e}$

(B) $\dfrac{1}{5e}$

(C) $\dfrac{1}{5}$

(D) $\dfrac{1}{e}$

(E) $\dfrac{1}{10}$

338. Find k if $\displaystyle\int_{0}^{9}\left(\dfrac{1}{\sqrt{x}} - k\right)dx = 12$.

(A) $-\dfrac{3}{2}$

(B) $-\dfrac{2}{3}$

(C) 0

(D) $\dfrac{2}{3}$

(E) $\dfrac{3}{2}$

339. If $f'(x) = g(x)$, express $\int_0^\pi g(2x)dx$ in terms of $f(x)$.

(A) $f(2\pi) - f(0)$

(B) $\dfrac{1}{2} f(2\pi)$

(C) $2f(\pi) - f(0)$

(D) $\dfrac{1}{2}[f(2\pi) - f(0)]$

(E) $2[f(2\pi) - f(0)]$

340. Evaluate $\displaystyle\int_{\ln 3}^{\ln 5} \dfrac{e^x}{e^x + 4} dx$.

(A) $\ln \dfrac{9}{7}$

(B) $\ln \dfrac{5}{7}$

(C) $\ln 2$

(D) $\dfrac{e^5}{e^5 + 4} - \dfrac{e^3}{e^3 + 4}$

(E) $e^5 - e^3$

341. Evaluate $\displaystyle\int_0^{\frac{\pi}{6}} \sqrt{\sin x} \cos x \, dx$.

(A) $\dfrac{2\sqrt{3}}{3}$

(B) $\dfrac{\sqrt{2}}{3}$

(C) $\dfrac{\sqrt{2}}{6}$

(D) $\dfrac{\sqrt{3}}{6}$

(E) $\dfrac{3\sqrt{2}}{8}$

342. If $G(x)$ is the antiderivative of $\ln(x)$ and $G(1) = 0$, find $G(2)$.

(A) $\ln 2$

(B) $2 \ln 2$

(C) $2 \ln 2 - 1$

(D) $2(\ln 2 - 1)$

(E) $2 \ln 2 + 1$

343. Evaluate $\displaystyle\int_0^{\frac{\pi}{2}} e^x \sin x \; dx$.

(A) $\dfrac{1}{2}$

(B) $\dfrac{1}{2}\left(e^{\frac{\pi}{2}} + 1 \right)$

(C) $e^{\frac{\pi}{2}} - 1$

(D) $e^{\frac{\pi}{2}} + 1$

(E) $\dfrac{1}{2} e^{\frac{\pi}{2}}$

344. Evaluate $\displaystyle\int_6^{10} \frac{1}{x^2 - 3x - 10}$.

(A) $\dfrac{1}{7} \ln \left| \dfrac{5}{96} \right|$

(B) $\dfrac{1}{7} \ln \left| \dfrac{10}{3} \right|$

(C) $\dfrac{1}{7} \ln \left| \dfrac{15}{2} \right|$

(D) $\dfrac{2}{5} \ln \left| \dfrac{10}{3} \right|$

(E) $\dfrac{2}{5} \ln \left| \dfrac{3}{10} \right|$

BC 345. Evaluate $\displaystyle\int_1^{\infty} \frac{1}{x^6} \, dx$.

(A) 0

(B) $\dfrac{1}{6}$

(C) $\dfrac{1}{5}$

(D) 1

(E) ∞

346. The marginal cost of producing x units of an item is $C'(x) = \dfrac{1}{4}x - 2$.

    (A) Find an expression for $C(x)$, assuming the cost of producing 0 units is \$2, such that $C(0) = 2$.

    (B) Find the value of x such that the average cost is a minimum.

    (C) Find the cost function $C(x)$ for producing 40 units.

347. There is an area A bounded by the curve $f(x) = x^2$ and the x-axis.

    (A) Use a Riemann sum to find area A on the interval from $x = 0$ to $x = 4$, using 4 subdivisions of equal length.

    (B) Find the area A on the interval from $x = 0$ to $x = 4$ using 8 subdivisions of equal length.

    (C) Now find area A by integrating $f(x)$ over the interval $(0, 4)$.

348. A virus population is growing at a rate of $P'(t) = 10t - 2\sqrt{t} + 100$ organisms per hour every t hours.

    (A) If the initial population, $P(0)$, is 500, what is the population after t hours?

    (B) What is the increase in population after 3 h, rounded to the nearest whole number?

    (C) What is the average increase in population per hour over the first 12 h?

349. A particle begins accelerating from a point 100 units along the x-axis and an initial velocity of $v(0) = 50$. The acceleration is given by $a(t) = 15\sqrt{t}$.

    (A) Find the equation of motion of the particle.

    (B) Find the position function of the particle.

    (C) What is the change in velocity between time $t = 0$ and $t = 5$?

    (D) What is the change in position from $t = 0$ to $t = 5$?

350. A pillar is 35 ft tall. The density of the pillar is given by $\rho(x) = \dfrac{1}{3\sqrt{x} + 1}$, where x is the distance from the ground ($x = 0$) in feet and the mass is measured in tons. Assume that the volume is constant over the length of the pillar.

    (A) What is the total mass of the pillar rounded to the nearest ton?

    (B) For what value of height, h, does the interval $[0, h]$ contain half the total mass of the pillar?

    (C) What is the mass of the uppermost 5 ft of the pillar, rounded to the nearest hundredth?

Areas and Volumes

351. If you were to use 3 midpoint rectangles of equal length to approximate the area under the curve of $f(x) = x^2 + 2$ from $x = 0$ to $x = 3$, how close would the approximation be to the exact area under the curve?

(A) $\dfrac{1}{2}$

(B) $\dfrac{1}{4}$

(C) $\dfrac{1}{8}$

(D) $\dfrac{3}{4}$

(E) $\dfrac{3}{8}$

352. Find the exact area of the region bounded by the graph of $f(x) = \sin x$, $g(x) = \cos x$, and the lines $x = 0$ to $x = \pi/2$.

(A) $\sqrt{2}$

(B) $2(\sqrt{3} + 2)$

(C) 1

(D) $\sqrt{2} - 2$

(E) $2(\sqrt{2} - 1)$

353. Find the exact area under the curve $f(x) = \tan x$ from $x = 0$ to $x = \dfrac{\pi}{3}$.

(A) $1 - \ln\left(\dfrac{1}{2}\right)$

(B) $1 + \dfrac{\sqrt{3}}{2}$

(C) 1

(D) $\ln 2$

(E) $2\sqrt{3}$

354. Solve for $b > 1$ when the area underneath the graph of $y = \dfrac{\ln x}{x}$ from $x = 1$ to $x = b$ is exactly 2.

(A) e^2

(B) e

(C) $\dfrac{1}{e - 2}$

(D) $\sqrt{e}$

(E) $\dfrac{e + 1}{e - 1}$

355. Find the area of the region bounded by the graphs of $f(x) = x^3$ and $g(x) = 3x^2 - 2x$ shown below.

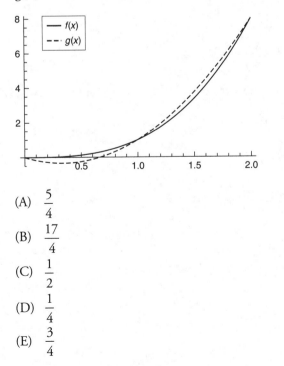

(A) $\dfrac{5}{4}$

(B) $\dfrac{17}{4}$

(C) $\dfrac{1}{2}$

(D) $\dfrac{1}{4}$

(E) $\dfrac{3}{4}$

356. Find the area of the region bounded by $y = e^{-x} + 2$, and the lines $y = -\dfrac{x}{2}$, $x = 0$, $x = 1$.

(A) $\dfrac{12}{4-e}$

(B) $\dfrac{12}{4e}$

(C) $\dfrac{9e-4}{4e}$

(D) $\dfrac{13e-4}{4e}$

(E) $\dfrac{13}{4e}$

357. Find the area of the region below the graph of $y = 2\sin(x)$ and above the line $y = 1$, from $x = 0$, and $x = 2\pi$.

(A) 1

(B) $\dfrac{1}{2}$

(C) $\dfrac{\sqrt{3}}{2}$

(D) $\sqrt{3} - \dfrac{5\pi}{6}$

(E) $2\left(\sqrt{3} - \dfrac{\pi}{3}\right)$

BC 358. Find the area under the curve $f(x) = 2x\ln x$ on the interval $[1, e]$.

(A) $\dfrac{1}{2}(e^2 - 1)$

(B) $\dfrac{1}{2}(e^2 + 1)$

(C) $\dfrac{e}{2} + 1$

(D) $2 + e^2$

(E) $1 + e^2$

359. Find the area under the curve $y = \cos^2(x)$ from $x = -\pi/6$ to $x = \pi/6$.

(A) $\dfrac{\pi}{12} - \dfrac{\sqrt{3}}{24}$

(B) $\dfrac{\pi}{6} - \dfrac{\sqrt{3}}{4}$

(C) $\sqrt{2} + \dfrac{\pi}{6}$

(D) $\dfrac{\pi}{6} + \dfrac{\sqrt{3}}{4}$

(E) $\dfrac{\pi}{12} + \dfrac{\sqrt{3}}{24}$

360. Find the area of the region bounded by $f(x) = 2e^{x/2}$, $y = 3$, and $x = 4$.

(A) $e^4 - \dfrac{\ln(3)}{\ln(2)}$

(B) $e^2 - \dfrac{3}{2}$

(C) $2e^2 - 3$

(D) $4e^2 + 6\ln\left(\dfrac{3}{2}\right) - 18$

(E) $e^2 + \dfrac{1}{2}\ln\left(\dfrac{3}{2}\right) - \dfrac{9}{2}$

361. Find the area under the curve of $f(x) = \dfrac{\sec x + \csc x}{1 + \tan x}$ from $x = \pi/6$ to $x = \pi/3$.

(A) 1.213
(B) 0.657
(C) 0.768
(D) 0.426
(E) 0.866

362. Find the area of the region bounded by the curves $f(x) = \ln(x)$ and $g(x) = 3 - 2\ln(x)$, and the line $x = e^2$.

(A) $3e$
(B) $9e$
(C) $-2 - 3e$
(D) $\dfrac{3e - 1}{e^2}$
(E) 1

363. Find the area of the region properly contained with $0 \le x \le \pi$ and bounded by the curves $y = \cos(x)$, $y = \cos(2x)$, and the y-axis.

(A) $\dfrac{3\sqrt{3}}{4}$

(B) $\sqrt{3}$

(C) 1

(D) $\dfrac{3\sqrt{2}}{4}$

(E) $\sqrt{2}$

364. Let $f(x) = x^2 + 1$. Let A represent the area under $y = f(x)$ from $x = 0$ to $x = a > 0$, B represent the area under $y = f(x)$ from $x = 0$ to $x = 2a$, and suppose $3A = B$. Solve for a.

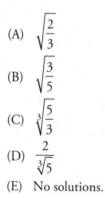

(A) $\sqrt{\dfrac{2}{3}}$

(B) $\sqrt{\dfrac{3}{5}}$

(C) $\sqrt[3]{\dfrac{5}{3}}$

(D) $\dfrac{2}{\sqrt[3]{5}}$

(E) No solutions.

365. Let $f(x) = \dfrac{1}{x}$. Let A represent the area under $y = f(x)$ from $x = 1$ to $x = a > 0$, B represent the area under $y = f(x)$ from $x = 1$ to $x = b$, and suppose $2A = B$. Write b in terms of a.

(A) $b = 2a$

(B) $b = a^2$

(C) $b = 4a$

(D) $b = a^4$

(E) $b = \dfrac{a^4}{2}$

366. Find the area of the region bounded from below by the line $y = 1$ and from above by the ellipse $x^2 + \dfrac{y^2}{4} = 1$.

(A) $\dfrac{\pi}{3} - \dfrac{\sqrt{2}}{2}$

(B) $\dfrac{2\pi}{3} + \dfrac{\sqrt{3}}{4}$

(C) $\dfrac{\pi}{6} + \dfrac{\sqrt{3}}{2}$

(D) $\dfrac{2\pi}{3} - \dfrac{\sqrt{3}}{2}$

(E) $\dfrac{\pi}{6} + \dfrac{\sqrt{2}}{2}$

367. If $g'(x) = f(x)$, which of the following represents the area under the curve $h(x) = xf(x^2)$ from $x = 0$ to $x = 3$?

(A) $2[g(3) - g(0)]$

(B) $\dfrac{1}{2}[g(3) - g(0)]$

(C) $2[f(9) - f(0)]$

(D) $\dfrac{1}{2}[g(9) - g(0)]$

(E) $\dfrac{1}{2}[f(9) - f(0)]$

368. (Use a calculator for this problem.) Find the area (rounded to three decimal places) of the region bounded by the curves $f(x) = e^x - 1$, $g(x) = \cos x$, and the y-axis.

(A) 0.195

(B) 0.449

(C) 1

(D) 0.274

(E) 0.343

BC 369. Find the area underneath the curve defined by the parametric equations $x = t + e^t$ and $y = 1 + e^t$, when $0 \le t \le \ln 2$.

(A) $\ln 2 + \dfrac{7}{2}$

(B) $2 + \ln\left(\dfrac{3}{2}\right)$

(C) $\ln 2 - \dfrac{7}{2}$

(D) $2 - \ln\left(\dfrac{3}{2}\right)$

(E) $\ln 2 - \dfrac{1}{2}$

BC 370. Find the area of the region located in the second quadrant and bounded by the y-axis and the curve defined by the parametric equations $x = t^2 + 4t$ and $y = 2 - t$.

(A) $\dfrac{-16}{3}$

(B) $\dfrac{-32}{3}$

(C) $\dfrac{64}{3}$

(D) $\dfrac{16}{3}$

(E) $\dfrac{32}{3}$

BC 371. Find the area of the region that lies inside the polar curve $r = \cos 2\theta$ and outside the circle $r = \dfrac{1}{2}$ when $-\pi/4 \le \theta \le \pi/4$.

(A) $\dfrac{\pi}{16} + \dfrac{\sqrt{3}}{24}$

(B) $\dfrac{\pi}{6} + \dfrac{\sqrt{3}}{12}$

(C) $\dfrac{\pi}{24} - \dfrac{\sqrt{3}}{16}$

(D) $\dfrac{\pi}{16} - \dfrac{\sqrt{3}}{24}$

(E) $\dfrac{\pi}{24} + \dfrac{\sqrt{3}}{16}$

BC 372. Find the area of the region located in the first and fourth quadrants and bounded by the polar curve $r = 3 + 2\sin\theta$.

(A) 3π

(B) $\dfrac{11\pi}{2}$

(C) $\dfrac{13\pi}{2}$

(D) $\dfrac{2\pi}{3}$

(E) $\dfrac{5\pi}{3}$

373. Consider a solid S whose base is the region enclosed by the curve $x = y^2$ and the line $x = 3$, and whose parallel cross sections perpendicular to the x-axis are squares. Find the volume of S.

(A) 6
(B) 9
(C) 18
(D) 27
(E) 54

374. Let S be a solid having as base the region in the first quadrant enclosed by the curve $xy = 3$ and the line $y = 4 - x$. Suppose further that parallel cross-sections of S perpendicular to the x-axis are rectangles having as base the vertical line connecting the graphs, and having height twice the base. Find the volume of S.

(A) 0.6
(B) 0.7
(C) 1.3
(D) 1.2
(E) 1.5

375. Consider a solid whose base is the region bounded by the curve $y = \log_2(x)$ and the lines $y = -1$ and $x = 2$. If parallel cross-sections perpendicular to the y-axis are regular triangles, what is the volume of the solid?

(A) 1.212
(B) 0.658
(C) 0.887
(D) 0.713
(E) 0.556

BC 376. Find the volume generated by revolving about the x-axis the region in the first quadrant bounded by the graph of $y = \ln x$ and the line $x = e$.

(A) $\pi(1 - \ln 2)$

(B) $2\pi(3 - e)$

(C) $\pi(\ln 3 - 1)$

(D) $\pi(3 - e)$

(E) $\pi(e - 2)$

377. Let D represent the region bounded by the unit circle centered at the origin. Find the volume of the solid obtained by revolving D about the line $x = 2$.

(A) $\dfrac{8\pi}{3} + 4\pi^2$

(B) $\dfrac{2\pi}{3} + \pi^2$

(C) $\pi\left(\dfrac{4}{3} + \sqrt{2}\right)$

(D) $\dfrac{8\pi}{3}$

(E) $\pi\left(\dfrac{4}{3} + \dfrac{\sqrt{3}}{2}\right)$

378. Find the volume of the solid obtained by revolving about the line $y = 1$ the region bounded by the curves $y = \tan x$ and $y = -\tan x$, the y-axis, and the line $x = \pi/4$.

(A) $\pi\left(1 + \dfrac{\pi}{2} - \ln 4\right)$

(B) $2\pi\left(1 - \dfrac{\pi}{4} + \ln 2\right)$

(C) $2\pi\left(1 - \dfrac{\pi}{2} + \ln 4\right)$

(D) $\pi\left(1 + \dfrac{3\pi}{4}\right)$

(E) $4\pi\left(1 - \dfrac{\pi}{4} + \ln\left(\dfrac{\pi}{4}\right)\right)$

379. Let $f(x) = \sec x$ if $-\pi/2 < x < \pi/2$. Find the volume of the solid obtained by rotating about the x-axis the region lying above the curve $y = f(x)$ and below the line $y = 2$.

(A) $4\pi\left(\dfrac{2\pi}{3} - \sqrt{3}\right)$

(B) $4\pi\left(\dfrac{2\pi}{3} + \sqrt{3}\right)$

(C) $2\pi\left(\dfrac{4\pi}{3} - \sqrt{3}\right)$

(D) $2\pi\left(\dfrac{4\pi}{3} + \sqrt{3}\right)$

(E) $\pi\left(\dfrac{2\pi}{3} + \dfrac{\sqrt{3}}{2}\right)$

380. What is the volume of the solid generated by rotating about $y = -1$ the region in the first quadrant bounded by the curves $y = 3 - x$ and $y = \dfrac{2}{x}$.

(A) $\dfrac{7\pi}{3}$

(B) $\pi\left[\dfrac{7}{3} - \pi \ln 4\right]$

(C) $\pi\left[\dfrac{9}{2} + \ln 3\right]$

(D) $\pi\left[\dfrac{10}{3} - 4\ln 2\right]$

(E) 3π

381. If a solid S has as base a triangular region with vertices $(0, 0)$, $(3, 0)$, and $(0, 1)$, and parallel cross-sections perpendicular to the x-axis are squares, what is the volume of S?

(A) $\dfrac{72}{5}$

(B) 7

(C) $\dfrac{45}{11}$

(D) 5

(E) $\dfrac{45}{7}$

382. Let $a > 0$, and consider the solid S obtained by revolving about the y-axis the region bounded by the curve $y = x$, the x-axis, and the line $x = a$. If the volume of S is known to be 18π, find a.

(A) 1

(B) $\dfrac{3}{2}$

(C) 2

(D) $\dfrac{5}{2}$

(E) 3

383. Find the volume of the solid obtained by rotating the region bounded by a right triangle with vertices at the origin, $(6, 0)$, and $(6, 3)$.

(A) 9π

(B) 18π

(C) 27π

(D) 36π

(E) 54π

BC 384. Find the area of the surface generated by revolving about the x-axis the curve defined by the parametric equations $x = 2\sin^2 t$ and $y = \sin(2t)$ when $0 \le t \le \pi/2$.

(A) π

(B) 2π

(C) 4π

(D) 6π

(E) 8π

BC 385. Find the area of the surface obtained by rotating about the x-axis the curve defined by the parametric equations $x = t^2 + 1$ and $y = t$, when $0 \le t \le 1$.

(A) $(1 - \sqrt{5})$

(B) $\dfrac{\pi}{8}\left(\dfrac{1 - \sqrt{5}}{\sqrt{5}} \right)$

(C) $\dfrac{\pi}{4}\left(\dfrac{1 + \sqrt{5}}{\sqrt{5}} \right)$

(D) $\dfrac{\pi}{4}(1 - \sqrt{5})$

(E) $\left(\dfrac{1 + \sqrt{5}}{\sqrt{5}} \right)$

BC 386. Find the surface area generated by rotating about the x-axis the curve defined by the parametric equations $x = 4e^{t/2}$ and $y = e^t - 4t$, when $0 \le t \le 1$.

(A) $\pi(e^2 - 16)$

(B) $\pi e(e + 8)$

(C) $2\pi(e^2 - 6e + 14)$

(D) $\pi(e^2 + 14e - 16)$

(E) $\pi(e^2 - 4e + 11)$

BC 387. Find the length of the polar curve $r = e^{\theta}$, when $0 \le \theta \le 2\pi$.

(A) $2(e^{\pi} - 2\pi)$

(B) $(e^{2\pi} - 1)$

(C) $\sqrt{3}(e^{\pi} - 2\pi)$

(D) $\sqrt{2}(e^{2\pi} - 1)$

(E) $\sqrt{2}(e^{4\pi} + 1)$

BC 388. Find the length to the polar curve $r = \sin\theta - \cos\theta$ when $0 \le \theta \le \pi$.

(A) $\pi - 1$

(B) $\sqrt{2\pi}$

(C) 2π

(D) $2\sqrt{2\pi} - 1$

(E) 4π

BC 389. The curve defined by the parametric equations $x = \cos t + t \sin t$ and $y = \sin t - t \cos t$, when $0 \le t \le a$ is known to have length $2\pi^2$. Find a.

(A) $\dfrac{\pi}{4}$

(B) $\dfrac{\pi}{2}$

(C) 2π

(D) 4π

(E) $\dfrac{4}{\pi^2}$

390. Let $f(t) = t^4 + 2t^2 - 3$, and consider the function $F(x) = \int_{-3}^{x} f(t)\, dt$. What are the x-value(s) for which $F(x)$ has a local minimum?

(A) 1

(B) −1

(C) $\pm\sqrt{3}$

(D) ± 1

(E) 1 and $\sqrt{3}$

391. Find the approximate area under the curve $y = e^x$ from $x = 0$ to $x = \ln 8$ using 3 trapezoids.

(A) $\dfrac{17 \ln 3}{3}$

(B) $\dfrac{19 \ln 2}{2}$

(C) $\dfrac{21 \ln 2}{2}$

(D) $\dfrac{11 \ln 3}{3}$

(E) $\dfrac{14 \ln 6}{3}$

BC 392. The acceleration of a particle moving in a plane is a vector function of time given by $a(t) = \langle \pi^2 \sin(\pi t), 6t \rangle$. If it is located at the origin $O = \langle 0,\, 0 \rangle$ when $t = 0$ and at $i + j$ when $t = 2$ (where $i = \langle 1,\, 0 \rangle$ and $j = \langle 0,\, 1 \rangle$ denote the standard unit vectors in two dimensions), find a formula for the position function $s(t)$, at any time t.

(A) $s(t) = \left(\dfrac{1}{2} - \sin(\pi t)\right) \cdot i + \left(t^3 - \dfrac{7}{2}\right) \cdot j$

(B) $s(t) = \sin(\pi t) \cdot i - t^3 \cdot j$

(C) $s(t) = \left(\dfrac{1}{2} + \sin(\pi t)\right) \cdot i - \left(t^3 + \dfrac{7}{2}\right) \cdot j$

(D) $s(t) = -\sin(\pi t) \cdot i + t^3 \cdot j$

(E) $s(t) = (1 - \sin(\pi t)) \cdot i + (t^3 - 7) \cdot j$

BC 393. Find the length of the vector curve $r(t) = \langle \ln | \sec t |, t \rangle$ from $t = 0$ to $t = \pi/4$.

(A) $\ln \left| 1 - \dfrac{1}{\sqrt{2}} \right|$

(B) $\ln \left| \sqrt{5} - 2 \right|$

(C) $\ln \left| \sqrt{2} + 1 \right|$

(D) $\ln \left| 2 - \sqrt{2} \right|$

(E) $\ln \left| \sqrt{3} - 1 \right|$

BC 394. Consider the vector curve $r(t) = \langle e^t, \sqrt{3} e^t \rangle$ from $t = 0$ to $t = k$. If the length of the curve is known to be 8, solve for k.

(A) $k = \ln 2$
(B) $k = 2$
(C) $k = \ln 3$
(D) $k = 3$
(E) $k = \ln 5$

395. If $g(x) = \displaystyle\int_0^x (\log_2 (t - 1) + \log_2 (t + 1))\, dt$, find all x-values at which the line tangent to the graph of $y = g(x)$ is horizontal.

(A) $\sqrt{2}$
(B) $\sqrt{3}$
(C) 2
(D) 3
(E) 4

Free Response Problems

396. Let $g(t) = \int_0^t f(x)\, dx$, and consider the graph of f shown below.

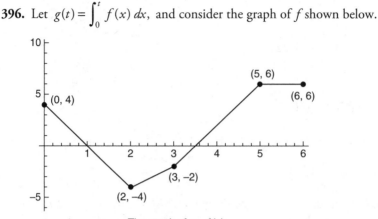

The graph of $y = f(x)$

(A) Evaluate $g(0)$, $g(2)$, and $g(6)$.
(B) On what interval(s) is g increasing (if any)?
(C) At what value(s) of t does g have a minimum value?
(D) On what interval(s) is g concave down?

397. Consider the two regions, R_1 and R_2, shown in the figure.

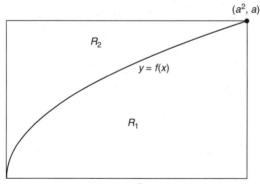

The rectangular region $[0, a^2] \times [0, a]$ cut in two by
the graph of $y = \sqrt{x}$.

(A) Is there a value of $a > 0$ which makes R_1 and R_2 have equal area? Justify your response.
(B) If the line $x = b$ divides the region R_1 into two regions of equal area, express b in terms of a.
(C) Express in terms of a the volume of the solid obtained by revolving the region R_1 about the y-axis.
(D) If R_2 is the base of a solid whose cross-sections perpendicular to the y-axis are squares, find the volume of the solid in terms of a.

398. Consider a triangle in the *xy*-plane with vertices at $A = (0, 1)$, $B = (2, 3)$, and $C = (3, 1)$. Let R denote the region that is bounded by the triangle shown in the figure.

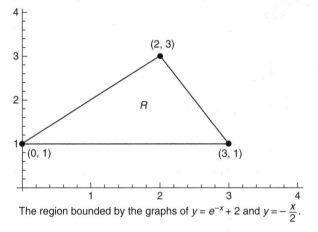

The region bounded by the graphs of $y = e^{-x} + 2$ and $y = -\dfrac{x}{2}$.

(A) Find the volume of the solid obtained by rotating R about the *x*-axis.

(B) Find the volume of the solid obtained by rotating R about the *y*-axis.

(C) Find the volume of the solid having R as its base while cross-sections perpendicular to the *x*-axis are squares.

399. Consider the region R in the first quadrant under the graph of $y = \cos x$ from $x = 0$ to $x = \pi/2$.

(A) Find the area of R.

(B) What is the volume of the solid obtained by rotating R about the *x*-axis?

(C) Suppose R is the surface of a concrete slab. If the depth of the concrete at x, where x is given in feet, is $d(x) = \sin x + 1$, find the volume (in cubic feet) of the concrete slab.

BC 400. Consider the curve defined parametrically by $x = \sin t - \cos t$ and $y = \sin t + \cos t$.

(A) Find the length of the curve from $\pi/4 \le t \le \pi/2$.

(B) Find the area bounded underneath the curve from $\pi/4 \le t \le \pi/2$.

(C) Find the area of the surface generated by revolving about the *x*-axis the parametric curve defined from $\pi/4 \le t \le \pi/2$.

More Applications of Definite Integrals

401. The graph of f is shown below. Find the average value of f on the interval $[-1, 6]$.

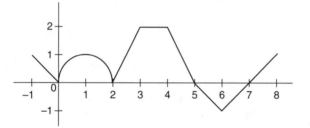

(A) $\dfrac{1}{2} + \dfrac{\pi}{16}$

(B) $\dfrac{5}{8} + \dfrac{\pi}{16}$

(C) $\dfrac{5}{7} + \dfrac{\pi}{14}$

(D) $\dfrac{4}{7} + \dfrac{\pi}{14}$

(E) $\dfrac{2}{3} + \dfrac{\pi}{16}$

402. The velocity of a particle moving on a line is given by the equation
$v(t) = 2t^2 - 14t - 5$. Find the average velocity from $t = 1$ to $t = 3$.

(A) $-\dfrac{5}{3}$

(B) 5

(C) $\dfrac{75}{4}$

(D) $\dfrac{73}{3}$

(E) $-\dfrac{73}{3}$

403. What is the average value of the function $y = 3\sin(2x) - 3\cos(2x)$ on the

interval $\left[-\dfrac{\pi}{2}, \dfrac{\pi}{6}\right]$?

(A) $\dfrac{9}{2\pi} + \dfrac{27\sqrt{3}}{4\pi}$

(B) $-\dfrac{9\pi}{4} - \dfrac{27\sqrt{3}}{4}$

(C) $-\dfrac{27}{4\pi} - \dfrac{9\sqrt{3}}{4\pi}$

(D) $\dfrac{9\pi}{8} - \dfrac{27\sqrt{3}}{4}$

(E) $9\pi + 27\sqrt{3}$

404. The graph of the acceleration function of a moving particle is shown below. On what intervals in [0, 20] does the particle have a positive change in velocity, and what is the total change in velocity of the particle?

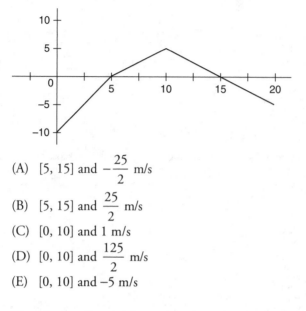

(A) [5, 15] and $-\dfrac{25}{2}$ m/s

(B) [5, 15] and $\dfrac{25}{2}$ m/s

(C) [0, 10] and 1 m/s

(D) [0, 10] and $\dfrac{125}{2}$ m/s

(E) [0, 10] and −5 m/s

405. The velocity of a particle moving along the number line is given by

$$v(t) = 7\cos\left(\frac{1}{7}t^2 + 1\right)$$ for $t \geq 0$. At $t = 1$ the position of the particle is 10.

Using your calculator, find the position of the particle when the velocity of the particle is equal to 0 for the first time.

(A) 1.635

(B) 3.14

(C) 15.129

(D) 11.635

(E) 5.129

406. A particle is moving along a straight line, and its acceleration is shown below. Assuming the particle started at rest, calculate the speed of the particle after 3 s.

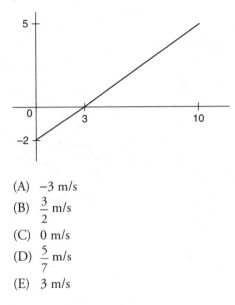

(A) −3 m/s

(B) $\dfrac{3}{2}$ m/s

(C) 0 m/s

(D) $\dfrac{5}{7}$ m/s

(E) 3 m/s

407. The graph of the velocity function of a moving car is shown below. In the interval [0, 15], calculate

(total distance of the car) − (total displacement of the car)

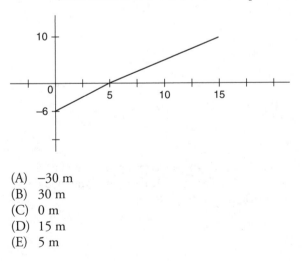

(A) −30 m

(B) 30 m

(C) 0 m

(D) 15 m

(E) 5 m

408. The velocity function of a moving particle on a coordinate line is given by $v(t) = t^2 - t - 2$ for $0 \leq t \leq 6$. Find (total distance traveled) − (total displacement) for this time period.

(A) $\dfrac{20}{3}$ m

(B) $-\dfrac{20}{3}$ m

(C) 0 m

(D) 46.5 m

(E) 3 m

409. The velocity function of a moving particle along a coordinate line is given by $v(t) = t^3 - 3t^2 - 2t + 4$, where velocity is given in meters per second. Using a calculator, for $0 \leq t \leq 5$ find

$$(\text{total displacement}) - (\text{total distance traveled})$$

(A) −12.5 m

(B) 12.5 m

(C) 0 m

(D) 15 m

(E) 32 m

410. The acceleration function of a moving particle on a coordinate line is $a(t) = -3$ with $v_0 = 9$ for $0 \leq t \leq 7$, where acceleration is caluclated in m/s². Find the total distance traveled by the particle.

(A) 0 m

(B) $-\dfrac{21}{2}$ m

(C) $-\dfrac{75}{2}$ m

(D) $\dfrac{21}{2}$ m

(E) $\dfrac{75}{2}$ m

411. The velocity function of a moving particle on a coordinate line is $v(t) = 2\sin(3t)$ for $-\pi \leq t \leq \pi$. Using a calculator, find the total distance traveled when the particle is moving to the right.

(A) −4

(B) 4

(C) π

(D) $\dfrac{\pi}{2}$

(E) 2π

412. The marginal profit of manufacturing and selling a flu vaccine is given by
$P'(x) = 1000 - 0.04x$, where x is the number of units of vaccine sold. How
much profit should the company expect if it sells 20,000 units of this vaccine?

(A) 12,000,000
(B) 1,200,000
(C) 120,000
(D) 12,000
(E) 1200

413. The marginal cost of producing x units of sneakers is $C'(x) = 5 + 0.4x$.
Find the cost of producing the first 100 pairs of sneakers.

(A) 2500
(B) 25,000
(C) 250
(D) 80
(E) 0

414. Graphs of the marginal cost and marginal revenue are shown below.
Estimate the profit for the first 50 units.

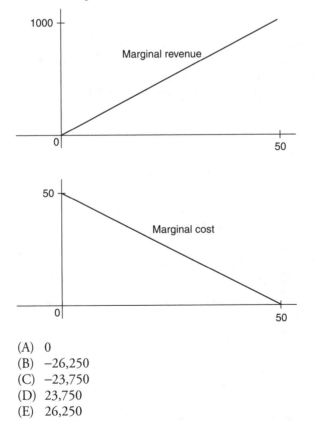

(A) 0
(B) −26,250
(C) −23,750
(D) 23,750
(E) 26,250

415. The temperature of a hot penny is dropping at the rate of $g(t) = 15e^{-.12t}$ for $0 \le t \le 5$, where $g(t)$ is measured in Fahrenheit and t in minutes. If the penny is initially 150°F, use your calculator to find the temperature of the penny to the nearest degree after 4 min.

  (A) 75°F
  (B) 100°F
  (C) 102°F
  (D) 47°F421
  (E) 197°F

416. On June 24, 2011 the changes in temperature of Des Plaines, IL, from 6 a.m. to 10 p.m. are represented by $f(t) = 3 \cos\left(\dfrac{t}{3}\right)$ degrees Fahrenheit, where t is number of hours elapsed after 6 a.m. Using your calculator, if at 6 a.m. the temperature is 75°F, find the temperature at 1 p.m.

  (A) 68.49
  (B) 81.51
  (C) 6.51
  (D) 75
  (E) 87.95

417. An approximation of the rate of change in air temperature at O'Hare Airport in Chicago is recorded through the day and shown below. If the air temperature is observed to be 60°F at 5 a.m., what number best approximates the temperature at 8 a.m.? The y-xais is measured in Fahrenheit.

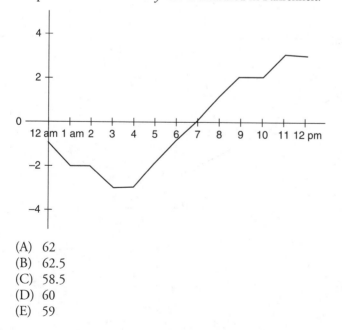

  (A) 62
  (B) 62.5
  (C) 58.5
  (D) 60
  (E) 59

418. Oil is leaking from a car at a rate of $f(t) = 12e^{-0.4t}$ oz/h, where t is measured in hours. Using your calculator, how many ounces will have leaked from the car after a 12-h period?

(A) 98.567
(B) 75.67
(C) 29.7531
(D) 12.34
(E) 4.98

419. Water is leaking from a tank at a rate of $f(t)$. The graph of $f(t)$ is shown below. Which of the following is the best approximation of the total amount of water leaked from the tank for $5 \le t \le 15$?

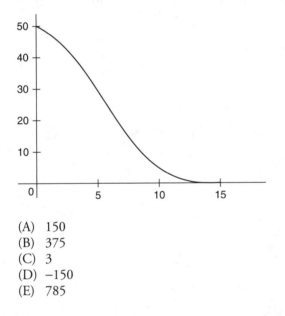

(A) 150
(B) 375
(C) 3
(D) −150
(E) 785

420. A full water tank begins to leak. The rate of leaking water can be approximated by $f(t) = 7e^{-0.1t} + 5$ gal/h, where t is measured in hours. After 12 h, the tank is half empty. Using your calculator, how much water did the full tank contain before it started leaking?

(A) 217.833
(B) 108.916
(C) 54.458
(D) 300
(E) 17

421. Between the months of February and July, sales of BBQs at Texas Depot increase at a rate which can be approximated by $f(t) = 30 + 20 \ln(1+t)$, where t is measured in weeks. Texas Depot sold no BBQs in January. Using your calculator, how many BBQs will Texas Depot sell by the end of the 7th week?

(A) 1123
(B) 600
(C) 100
(D) 403
(E) 75

422. A colony of bacteria is growing at a rate which can be approximated by $f(t) = 100 + 250 \ln(3 + t)$, where t is in hours. How much will the bacteria population increase between the 6th and 8th hour?

(A) 200
(B) 65
(C) 2467
(D) 945
(E) 1350

423. The rate of hybrid car production in America between 2001 and 2011 is shown below. Which number best approximates how much production increased between 2008 and 2011?

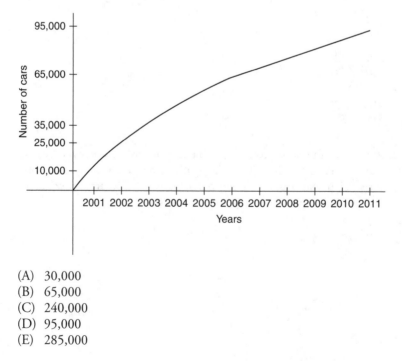

(A) 30,000
(B) 65,000
(C) 240,000
(D) 95,000
(E) 285,000

424. The price of stock in Abercrombie and Fitch on Febrary 4th, 2011 from 9 a.m. to 3 p.m. is approximated by the function $f(t) = \dfrac{x^2}{4} + 2$, where $t =$ number of hours after 9 a.m., when the stock exchange opens. Is there a time during the day when the value of the stock is equal to its average value for the entire day? If yes, at what time?

(A) 12:28 p.m.

(B) 1:34 p.m.

(C) 12:47 p.m.

(D) 11:52 a.m.

(E) At no time during the day

425. Given $f(x) = \sqrt{2x - 3}$ verify the hypothesis of the mean value theorem for f on [2, 8] and find the value of c as indicated in the theorem.

(A) ≈ -1.7

(B) ≈ 2.3

(C) ≈ 4.75

(D) ≈ 10

(E) ≈ 25

426. Plutonium has a half-life of 8645 years. Using your calculator, if there are initially 20 g of plutonium, how many grams are left after 1000 years?

(A) ≈ 45

(B) ≈ 18

(C) ≈ 35

(D) ≈ 15

(E) ≈ 12

427. The amount of bacteria in a petri dish increases at a rate proportional to the amount of bacteria present. An initial amount of bacteria is placed in the dish, and shortly after they begin to multiply. Using your calculator, if there are 200 bacteria after one day in the dish, and 600 bacteria after the 3 days in the dish, how many bacteria are in the petri dish after the 7 days?

(A) 5400

(B) 400

(C) 54

(D) 500

(E) 540

428. The temperature of your coffee decreases according to the equation $\frac{dy}{dt} = ky$ with t measured in minutes. If after 5 min the temperature decreases by 70%, then $k = ?$

(A) $\dfrac{\ln{(3)}}{5 \ln{(10)}}$

(B) $\dfrac{\ln\left(\dfrac{3}{10}\right)}{5}$

(C) $\dfrac{\ln{(3)} - \ln{(10)}}{5}$

(D) Both B and C

(E) Both A and B

429. The rate of growth of the wolf population in Yosemite National Park is proportional to the population. The wolf population increased by 11% between 2002 and 2011. What is the constant of proportionality?

(A) 3.6

(B) 0.2

(C) 11.985

(D) 0.0116

(E) −1.2

430. Write an equation for the curve that passes through the point (2, 11) and whose slope is $\dfrac{4xy}{2x^2 + 3}$.

(A) $y = -(2x^2 + 3)$

(B) $y = (2x^2 + 3)$

(C) $y = (3x^2 + 2)$

(D) $y^2 = (2x^2 + 3)$

(E) $y^2 = (2x^2 + 3)^2$

431. If $\frac{dy}{dx} = -x \cos(x^2)$ and $y = 2$ when $x = 0$, then a solution to the differential equation is:

(A) $y = -\frac{1}{2} \sin(x^2)$

(B) $y = -\frac{1}{2} (\cos(x))^2 + 2$

(C) $y = -\frac{1}{2} (\sin(x))^2 + 2$

(D) $y = -\frac{1}{2} \cos(x^2) + 2$

(E) $y = -\frac{1}{2} \sin(x^2) + 2$

432. If $\frac{dy}{dx} = 5x^4 y^2$ and $y = 1$ when $x = 1$, then determine y when $x = -1$?

(A) 1

(B) −1

(C) $\frac{1}{3}$

(D) $\frac{-1}{3}$

(E) 0

433. If $\frac{d^2 y}{dx^2} = 4x - 5$ and when $x = 0$, $y' = 3$ and $y = 4$, find a solution of the differential equation.

(A) $y = \frac{2}{3}x^3 - \frac{5}{2}x^2 + 3x + 4$

(B) $y = 2x^2 - 5x + 3$

(C) $y = \frac{3}{2}x^3 - \frac{2}{5}x^2 + 3x + 4$

(D) $y = 6x^3 - 10x^2 + 3x + 4$

(E) $y = (4x - 5)^3$

434. The figure below shows a slope field for one of the differential equations given below. Identify the equation.

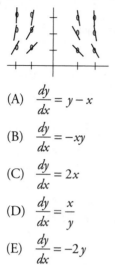

(A) $\dfrac{dy}{dx} = y - x$

(B) $\dfrac{dy}{dx} = -xy$

(C) $\dfrac{dy}{dx} = 2x$

(D) $\dfrac{dy}{dx} = \dfrac{x}{y}$

(E) $\dfrac{dy}{dx} = -2y$

435. The figure below shows a slope field for one of the differential equations given below. Identify the equation.

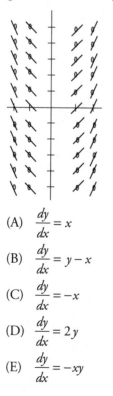

(A) $\dfrac{dy}{dx} = x$

(B) $\dfrac{dy}{dx} = y - x$

(C) $\dfrac{dy}{dx} = -x$

(D) $\dfrac{dy}{dx} = 2y$

(E) $\dfrac{dy}{dx} = -xy$

436. What family of functions has a slope field given by the differential

equation $\dfrac{dy}{dx} = \dfrac{x}{y}$?

(A) lines
(B) ellipses
(C) hyperbolas
(D) circles
(E) parabolas

437. If $\dfrac{dy}{dx} = y - x$, identify the slope field.

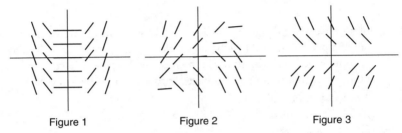

Figure 1   Figure 2   Figure 3

(A) Figure 1
(B) Figure 2
(C) Figure 3
(D) all of the above
(E) none of the above

438. The rabbit population of a particular warren in Cook County Forest
Preserve was 17 rabbits in 2001 and 35 rabbits in 2008. Find a logistic
model for the growth of the population, assuming a carrying capacity of
75 rabbits. Use the model to predict the population in 2013.

(A) 69
(B) 75
(C) 38
(D) 57
(E) 157

439. The spread of Asian Beetles through a grove of 60 trees is modeled by

$$\frac{dP}{dt} = .71P\left(1 - \frac{P}{60}\right)$$

On day zero, one tree is infected. Find the logistic model for the population of infected trees at time t, and use it to predict when half the trees will be infected.

(A) 2 days
(B) 4 days
(C) 6 days
(D) 8 days
(E) 10 days

440. John Hersey High School, which has 1532 students, is circulating a rumor that 3rd period on Friday will be canceled for a fire drill. On Monday, 7 people have heard the rumor. On Tuesday, 84 people have heard the rumor. How many people will have heard the rumor by Thursday?

(A) 1384 students
(B) 1178 students
(C) 785 students
(D) 598 students
(E) 345 students

441. A college dorm that houses 300 students experiences an outbreak of measles. The Health Center recognizes the outbreak when 4 students are diagnosed on the same day. Residents are quarantined to restrict this infection to one building. After 7 days, 17 students are sick with measles. Use a logistic model to describe the course of the infection and predict the number of students infected after 14 days.

(A) 34 students
(B) 44 students
(C) 54 students
(D) 64 students
(E) 74 students

442. Given $\frac{dy}{dt} = 2\sin(4\pi t)$ and $y(1) = 2$, approximate $y(3)$ using five steps.

(A) 8
(B) −17
(C) −2
(D) 17
(E) 2

443. Use Euler's method and a step size of $\Delta x = 0.2$ to compute $y(1)$ if $y(x)$ is the solution of the differential equation

$$\frac{dy}{dx} + 4x^3 y = 2x^3$$

with initial condition $y(0) = 2$.

(A) 4.8
(B) 1.19
(C) −12.45
(D) 5.67
(E) −2.4

444. Approximate $P(4)$ given

$$\frac{dP}{dt} = -.2P\left(1 - \frac{3P}{5}\right)$$

with initial condition $P(2) = 3$. Use three steps.

(A) 4.4
(B) 9.6
(C) 27.2
(D) −5.4
(E) 1.2

445. The velocity of a ball thrown from a 50-ft cliff is shown below. Use Euler's method to approximate the distance the ball is from the ground after 2 s. Use a step size of 0.5. (Graph is not drawn to scale.)

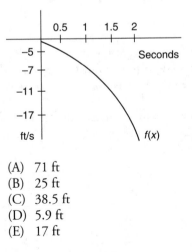

(A) 71 ft
(B) 25 ft
(C) 38.5 ft
(D) 5.9 ft
(E) 17 ft

446. The change in temperature of a greenhouse from 7 p.m. to 7 a.m. is given by the function:

$$f(t) = -3 \sin\left(\frac{t}{3}\right)$$

where $f(t)$ is measured in Fahrenheit and t is the number of hours after 7 p.m.

(A) If at 7 p.m. the temperature is 105°F, find the temperature of the greenhouse at 2 a.m.

(B) Write an expression to represent the temperature of the greenhouse at time t, where t is between 7 p.m. and 7 a.m.

(C) Find the average change in temperature of the greenhouse between 7 p.m. and 7 a.m. to the nearest 10th of a degree.

(D) Is there a point during the night when the change in temperature of the greenhouse is the average change in temperature? If yes, state the time. Either way, show all work to justify your answer.

447. Consider the differential equation given by $\dfrac{dy}{dx} = \dfrac{4xy}{5}$.

(A) On the axis provided below, sketch a slope field for the given differential equation at the points indicated.

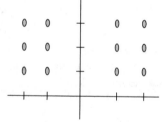

(B) Let $y = f(x)$ be the particular solution to the given differential equation with initial condition $f(0) = 5$. Use Euler's method and a step size of 0.1 to approximate $f(0.3)$. Show the work that leads to your answer.

(C) Find the particular solution $y = f(x)$ to the given differential equation with initial condition $f(0) = 2$.

(D) Use your solution above to find $f(0.3)$.

448. This problem is solved without a calculator. The slope of a function f at any given point (x, y) is $\dfrac{2y}{3x^2}$. The point $(3, 4)$ is on the graph of f.

(A) Write an equation of the tangent line to the graph of f at $x = 3$.

(B) Use the tangent line in part (A) to approximate $f(5)$.

(C) Solve the seperable differential equation $\dfrac{dy}{dx} = \dfrac{2y}{3x^2}$ with initial condition $f(3) = 4$.

(D) Use the solution in part (C) to find $f(5)$.

449. A water barrel containing 200 gal of water is punctured and begins to leak at a rate of $15 \sin\left(\dfrac{\pi t}{30}\right)$ gal/min, where t is measured in minutes and $0 \le t \le 15$.

(A) How much water to the nearest gallon leaked out after $t = 7$ min?

(B) What is the average amount of water leaked out per minute from $t = 0$ to $t = 7$ to the nearest gallon?

(C) Write an expression for $f(t)$ to represent the total amount of water in the barrel at time t, where $0 \le t \le 15$.

(D) At what value of t to the nearest minute will there be 50 gal of water remaining in the barrel?

450. A butterfly population is modeled by a function P that satisfies the logistic differential equation:

$$\frac{dP}{dt} = \frac{1}{7}P\left(1 - \frac{P}{21}\right)$$

(A) If $P(0) = 15$, what is $\lim_{t \to \infty} P(t)$?

(B) For what value of P is the population growing fastest?

(C) A different population is modeled by the function Q that satisfies the separable differential equation:

$$\frac{dQ}{dt} = \frac{1}{7}Q\left(1 - \frac{t}{21}\right)$$

Find $Q(t)$ if $Q(0) = 10$.

(D) For the function found in part (C), what is $\lim_{t \to \infty} Q(t)$?

Series (for Calculus BC Students Only)

451. Which of the following statements concerning the sequence $\{a_n\} = \dfrac{n}{2n^2 - 3}$ is true?

(A) Both $\{a_n\}$ and $\sum\limits_{n=1}^{\infty} a_n$ are convergent.

(B) $\{a_n\}$ is convergent, but $\sum\limits_{n=1}^{\infty} a_n$ is divergent.

(C) $\{a_n\}$ is divergent, but $\sum\limits_{n=1}^{\infty} a_n$ is convergent.

(D) Both $\{a_n\}$ and $\sum\limits_{n=1}^{\infty} a_n$ are divergent.

452. If the nth partial sum of a series $\sum\limits_{i=1}^{\infty} a_i$ is

$$S_n = \frac{3n+1}{2n-5}$$

find a_5.

(A) $\dfrac{3}{2}$

(B) $\dfrac{17}{5}$

(C) $-\dfrac{17}{15}$

(D) $\dfrac{16}{5}$

(E) $-\dfrac{16}{5}$

453. If the nth partial sum of a series $\sum_{i=1}^{\infty} a_i$ is

$$s_n = \frac{(\ln n)^2}{n}$$

find $\sum_{i=1}^{\infty} a_i$.

(A) $-\infty$

(B) 0

(C) 1

(D) e^2

(E) ∞

454. Find the sum of the infinite series

$$4 + \frac{4}{3} + \frac{4}{9} + \frac{4}{27} + \cdots$$

(A) $\dfrac{85}{3}$

(B) 24

(C) $\dfrac{20}{3}$

(D) 6

(E) 5

455. Express $1.3\overline{12}$ as a ratio of two integers.

(A) $\dfrac{521}{330}$

(B) $\dfrac{417}{495}$

(C) $\dfrac{1049}{990}$

(D) $\dfrac{559}{495}$

(E) $\dfrac{433}{330}$

456. Is the series $s = \sum_{n=1}^{\infty} 3^{2n} 2^{3-3n}$ convergent or divergent? If convergent, what is the sum?

(A) convergent; sum $= -72$
(B) convergent; sum $= 72$
(C) convergent; sum $= 8$
(D) convergent; sum $= 9$
(E) divergent

457. Find the sum of the convergent series $\sum_{i=1}^{\infty} \left(\dfrac{1}{i} - \dfrac{1}{i+2} \right)$.

(A) $\dfrac{3}{2}$

(B) 1

(C) 2

(D) $\dfrac{5}{3}$

(E) $\dfrac{7}{4}$

458. If $\sum_{n=1}^{\infty} a_n = 3$ and $\sum_{n=1}^{\infty} b_n = 7$, find the sum of the infinite series $\sum_{n=1}^{\infty} (6a_n - 2b_n)$.

(A) 4
(B) 10
(C) 21
(D) 32
(E) 36

459. Find the values of c so that the series $\sum_{n=1}^{\infty} \dfrac{1}{n(\ln n)^c}$ is convergent.

(A) $c \leq 0$
(B) $c > 0$
(C) $c \leq 1$
(D) $c > 1$
(E) No such c exists.

460. Determine whether $\sum_{n=2}^{\infty} \dfrac{5}{n^2 - n}$ converges or diverges.

(A) The series converges.
(B) The series diverges.

461. Determine whether $\sum_{n=1}^{\infty} \dfrac{3n+7}{5n-2}$ converges or diverges.

(A) The series converges.
(B) The series diverges.

462. Determine whether $\sum_{n=1}^{\infty} \dfrac{\ln n}{n^2}$ converges or diverges.

(A) The series converges.
(B) The series diverges.

463. Determine whether $\sum_{n=1}^{\infty} \dfrac{2 + \sin n}{n^2}$ converges or diverges.

(A) The series converges.
(B) The series diverges.

464. Determine whether $\sum_{n=1}^{\infty} \sin\left(\dfrac{1}{n^2}\right)$ converges or diverges.

(A) The series converges.
(B) The series diverges.

465. Determine whether $\sum_{n=1}^{\infty} \dfrac{n^n}{n!}$ converges or diverges.

(A) The series converges.
(B) The series diverges.

466. Determine whether $\sum_{n=2}^{\infty} (-1)^n \dfrac{1}{\ln n}$ converges or diverges.

(A) The series converges.
(B) The series diverges.

467. Determine whether $\sum_{n=1}^{\infty} (-1)^{n-1} \dfrac{e^{1/n}}{\sqrt{n}}$ converges or diverges.

(A) The series converges.
(B) The series diverges.

468. Determine whether the series $\sum_{n=1}^{\infty} (-1)^n \dfrac{\cos(\pi n)}{n}$ is absolutely convergent, conditionally convergent, or divergent.

(A) absolutely convergent
(B) conditionally convergent
(C) divergent

469. The series $\displaystyle\sum_{n=0}^{\infty} a_n$ is defined recursively by the equations

$$a_0 = 0 \qquad \text{and} \qquad a_{n+1} = \left(\frac{\sin n + n}{2n+1}\right) a_n$$

Determine whether $\displaystyle\sum_{n=0}^{\infty} a_n$ converges or diverges.

(A) The series converges.
(B) The series diverges.

470. Determine whether $\displaystyle\sum_{n=0}^{\infty} n^2 e^{-2n}$ converges or diverges.

(A) The series converges.
(B) The series diverges.

471. Determine whether the series

$$\frac{1}{2} - \frac{2}{5} + \frac{3}{10} - \frac{4}{17} + \cdots$$

is absolutely convergent, conditionally convergent, or divergent.

(A) The series converges absolutely.
(B) The series converges conditionally.
(C) The series diverges.

472. Determine whether the series $\displaystyle\sum_{n=1}^{\infty} (-1)^{n+1}\left(\frac{n}{2n+1}\right)^n$ is absolutely convergent, conditionally convergent, or divergent.

(A) The series converges absolutely.
(B) The series converges conditionally.
(C) The series diverges.

473. Approximate the sum of the alternating series $s = \displaystyle\sum_{n=1}^{\infty} \frac{(-2)^n n}{8^n}$ correct to three decimal places.

(A) $s \approx -0.840$
(B) $s \approx 0.005$
(C) $s \approx -0.160$
(D) $s \approx 0.572$
(E) $s \approx -1.080$

474. For what values of x is the series $\displaystyle\sum_{n=1}^{\infty} \frac{x^n}{n2^n}$ convergent?

(A) $x \in (-1, 1]$
(B) $x \in [-1, 2)$
(C) $x \in (-\infty, \infty)$
(D) $x \in (-2, 2)$
(E) $x \in [-2, 2)$

475. Find the radius of convergence for the power series $\displaystyle\sum_{n=0}^{\infty} n^3 (x-4)^n$.

(A) 0
(B) 1
(C) 2
(D) 3
(E) ∞

476. Find the interval of convergence for the power series $\displaystyle\sum_{n=3}^{\infty} \frac{x^{2n}}{(\ln n)^n}$.

(A) $x = 0$
(B) $x \in [-1, 1)$
(C) $x \in (-\infty, \infty)$
(D) $x \in (-\sqrt{e}, \sqrt{e})$
(E) $x \in [-\sqrt{e}, \sqrt{e})$

477. Find a power series representation for $\dfrac{x}{2+3x}$ and determine the interval of convergence I.

(A) $\displaystyle\sum_{n=1}^{\infty} \frac{(-3)^n}{2^{n+1}} x^{n+1}$ with $I = \left(-\frac{2}{3}, \frac{2}{3}\right)$

(B) $\displaystyle\sum_{n=0}^{\infty} \left(\frac{4}{3}\right)^n x^n$ with $I = \left(-\frac{3}{4}, \frac{3}{4}\right)$

(C) $\displaystyle\sum_{n=0}^{\infty} \frac{4^n}{3^{n+1}} x^{n+1}$ with $I = \left(-\frac{3}{4}, \frac{3}{4}\right)$

(D) $\displaystyle\sum_{n=1}^{\infty} \left(\frac{3}{2}\right)^n x^n$ with $I = \left[-\frac{2}{3}, \frac{2}{3}\right)$

(E) $\displaystyle\sum_{n=1}^{\infty} \frac{(-3)^{n-1}}{2^n} x^n$ with $I = \left(-\frac{2}{3}, \frac{2}{3}\right)$

478. Find a power series representation for $f(x) = \ln(1 - 2x)$ and determine the radius of convergence r.

(A) $-\sum_{n=1}^{\infty} \frac{2^n}{n} x^n; \ r = \frac{1}{2}$

(B) $\frac{1}{2} \sum_{n=1}^{\infty} x^n; \ r = 1$

(C) $-\sum_{n=1}^{\infty} \frac{2^n}{n} x^{n+1}; \ r = \frac{1}{2}$

(D) $\sum_{n=0}^{\infty} \frac{4^{n+2}}{n+1} x^n; \ r = \frac{1}{4}$

(E) $-\sum_{n=0}^{\infty} 2^{n+1} x^n; \ r = \frac{1}{2}$

479. Find a power series representation for $f(x) = \dfrac{1}{(1+x)^2}$ and determine the radius of convergence r.

(A) $\sum_{n=1}^{\infty} \frac{(-1)^n}{n} x^n; \ r = 1$

(B) $\sum_{n=1}^{\infty} \frac{1}{n+1} x^n; \ r = 1$

(C) $\sum_{n=1}^{\infty} (-1)^n n x^n; \ r = \frac{1}{2}$

(D) $\sum_{n=0}^{\infty} (-1)^n (n+1) x^n; \ r = 1$

(E) $\sum_{n=0}^{\infty} (-1)^n n x^n; \ r = \frac{1}{2}$

480. Find the interval of convergence for the power series $\sum_{n=1}^{\infty} n^n x^{2n}$.

(A) $x = 0$

(B) $x \in (-\sqrt{2}, \sqrt{2})$

(C) $x \in (-2, 2)$

(D) $x \in (-1, 1)$

(E) $x \in (-\infty, \infty)$

481. Let $s = 10 + 4 + \dfrac{8}{5} + \dfrac{16}{25} + \dfrac{32}{125} + \cdots$. Find the sum of the first 10 terms.

(A) 20.132

(B) 16.667

(C) $20.\overline{8}$

(D) 16.274

(E) $18.\overline{6}$

482. Find the Maclaurin series of the function $f(x) = x^2 e^{2x}$ and its radius of convergence r.

(A) $\sum_{n=0}^{\infty} \dfrac{2^{n+1}}{(n+1)!} x^n; \ r = \infty$

(B) $\sum_{n=0}^{\infty} \dfrac{4^n}{(n+1)!} x^{n+2}; \ r = 1$

(C) $\sum_{n=0}^{\infty} \dfrac{2^n}{n!} x^{n+2}; \ r = \infty$

(D) $\sum_{n=0}^{\infty} \dfrac{2^n}{n!} x^n; \ r = \dfrac{1}{2}$

(E) $2\sum_{n=0}^{\infty} \dfrac{x^n}{n!}; \ r = \infty$

483. Find the degree 2 Maclaurin polynomial which approximates
$f(x) = \ln\left(\dfrac{x+1}{(x+2)^2}\right)$.

(A) $-2\ln 2 + x - 2x^2$

(B) $\ln 2 - \dfrac{1}{4}x^2$

(C) $2 - x^2$

(D) $\ln 2 + x - 2x^2$

(E) $-2\ln 2 - \dfrac{1}{4}x^2$

484. If a function f is approximated by the third order Taylor series
$4 - 3(x-2) + 2(x-2)^2 - 7(x-3)^3$ centered at $x = 2$, what is $f'''(2)$?

(A) -42

(B) -21

(C) $-\dfrac{7}{6}$

(D) $\dfrac{7}{6}$

(E) $\dfrac{13}{2}$

485. Find a Maclaurin series for $f(x) = \cos^2(x)$.

(A) $\dfrac{1}{2} + \displaystyle\sum_{n=0}^{\infty} \dfrac{(-1)^n 2^{2n}}{(2n)!} x^{2n-1}$

(B) $\displaystyle\sum_{n=0}^{\infty} \dfrac{(-1)^n}{(2n)!} x^{2n}$

(C) $\displaystyle\sum_{n=0}^{\infty} \dfrac{2^{2n-1}}{(2n)!} x^{n}$

(D) $\dfrac{1}{2} + \displaystyle\sum_{n=0}^{\infty} \dfrac{(-1)^n 2^{2n-1}}{(2n)!} x^{2n}$

(E) $\dfrac{1}{2} + \displaystyle\sum_{n=0}^{\infty} \dfrac{1}{(2n-1)!} x^{2n}$

486. Use Maclaurin series to approximate the integral $\int_0^1 \dfrac{e^{-x}-1}{x}\,dx$ to three decimal places.

(A) 1.387
(B) −0.796
(C) −2.558
(D) −0.288
(E) 1.965

487. Find the Taylor series for $f(x) = \dfrac{1}{\sqrt{x}}$ centered at $x = 4$.

(A) $\displaystyle\sum_{n=0}^{\infty} (-1)^n \frac{1 \cdot 3 \cdot 5 \cdots (2n-1)}{2^{3n+1} n!}(x-4)^n$

(B) $\displaystyle\sum_{n=0}^{\infty} (-1)^n \frac{1 \cdot 3 \cdot 5 \cdots (2n+1)}{4^{n+1}}(x-4)^n$

(C) $\displaystyle\sum_{n=0}^{\infty} \frac{1}{2^{2n+1} n} x^n$

(D) $\displaystyle\sum_{n=1}^{\infty} (-1)^n \frac{2 \cdot 4 \cdot 6 \cdots 2n}{2^{3n+1} n!}(x-4)^n$

(E) $\displaystyle\sum_{n=0}^{\infty} (-1)^n \frac{(2n)!}{2^{3n+1} n!}(x-4)^n$

488. Find the Taylor series for $f(x) = \cos(\pi x)$. centered at $x = 3$.

(A) $\displaystyle\sum_{n=0}^{\infty} \frac{(-1)^n \pi^{2n-2}}{(2n)!}(x-1)^{2n}$

(B) $\displaystyle\sum_{n=0}^{\infty} \frac{(-1)^{n+1} \pi^{2n}}{(2n)!}(x-1)^{2n}$

(C) $\displaystyle\sum_{n=0}^{\infty} \frac{\pi^{2n-1}}{(2n-1)!} x^{2n-1}$

(D) $\displaystyle\sum_{n=0}^{\infty} \frac{(-1)^{n+1}}{n! \pi^{2n}}(x-1)^n$

(E) $\displaystyle -\sum_{n=0}^{\infty} \frac{1}{(2n)! \pi^{2n}}(x-1)^{2n}$

489. Find the sum of the infinite series $\sum_{n=0}^{\infty} \dfrac{(-1)^n 2^n}{3^n n!}$.

(A) $e^{3/2}$

(B) 3

(C) $-e^{3/2}$

(D) $-\dfrac{2}{5}$

(E) $e^{-2/3}$

490. Find the first three nonzero terms in the Maclaurin series of $f(x) = \dfrac{e^x}{1-x}$.

(A) $x + \dfrac{1}{2}x^2 + \dfrac{5}{6}x^3 + \cdots$

(B) $1 + \dfrac{1}{2}x + \dfrac{3}{2}x^2 + \cdots$

(C) $1 + 2x + \dfrac{5}{2}x^2 + \cdots$

(D) $1 + x + \dfrac{3}{2}x^2 + \cdots$

(E) $x + x^2 + \dfrac{5}{6}x^3 + \cdots$

491. Approximate $\cos(9°)$ accurate to three decimal places.

(A) 0.874

(B) 0.891

(C) 0.951

(D) 0.982

(E) 0.987

492. Find the sum of the infinite series $\dfrac{\pi}{3} - \dfrac{(\pi/3)^3}{3!} + \dfrac{(\pi/3)^5}{5!} - \dfrac{(\pi/3)^7}{7!} + \cdots$

(A) $\dfrac{1}{2}$

(B) $\dfrac{\sqrt{3}}{2}$

(C) ∞

(D) $-\dfrac{\sqrt{3}}{2}$

(E) $-\dfrac{1}{2}$

493. Evaluate the indefinite integral $\displaystyle\int \frac{\cos(x^3)-1}{x}\,dx$ as an infinite series.

(A) $\displaystyle\sum_{n=0}^{\infty} \frac{(-1)^n}{3n \cdot n!} x^{3n} + c$

(B) $\displaystyle\sum_{n=1}^{\infty} \frac{1}{6(n-1)(2n)!} x^{6n-1} + c$

(C) $\displaystyle\sum_{n=1}^{\infty} \frac{(-1)^{n-1}}{6n(2n-1)!} x^{6n-3} + c$

(D) $\displaystyle\sum_{n=0}^{\infty} \frac{1}{6n(2n)!} x^{6n} + c$

(E) $\displaystyle\sum_{n=1}^{\infty} \frac{(-1)^n}{6n(2n)!} x^{6n} + c$

494. Use the third-degree Taylor polynomial $T_3(x)$ to approximate $\ln(2)$.

(A) $\dfrac{2}{3}$

(B) $\dfrac{8}{11}$

(C) $\dfrac{7}{12}$

(D) $\dfrac{9}{13}$

(E) $\dfrac{13}{20}$

495. Which of the following converge to 3?

I. $\displaystyle\sum_{n=0}^{\infty} \frac{2}{3^n}$

II. $1 + \ln 3 + \dfrac{(\ln 3)^2}{2} + \dfrac{(\ln 3)^3}{6} + \dfrac{(\ln 3)^4}{24} + \cdots$

III. $\displaystyle\sum_{n=0}^{\infty} \frac{3n+1}{n-8}$

(A) I only
(B) II only
(C) III only
(D) I and II only
(E) I and III only

Free Response Problems

496. (A) Show that the series $\sum_{n=1}^{\infty} \dfrac{n^n}{(3n)!}$ is convergent.

(B) Use part A to evaluate the limit $\lim_{n \to \infty} \dfrac{n^n}{(3n)!}$.

497. (A) Approximate $f(x) = \dfrac{1}{x^2}$ by $T_3(x)$, the Taylor polynomial with degree 3 centered at 1.

(B) Use Taylor's inequality to estimate the accuracy of the approximation $f(x) \approx T_3(x)$ when x lies in $0.8 \le x \le 1.2$.

498. Suppose the Nth partial sum of a series $\sum_{n=0}^{\infty} a_n$ is

$$s_N = N \tan\left(\frac{\pi}{N}\right)$$

(A) Find a_4.

(B) Find $\sum_{n=0}^{\infty} a_n$.

499. Suppose that $\sum_{n=0}^{\infty} c_n x^n$ converges when $x = -3$ and diverges when $x = 5$.

Using this information, determine whether the following series converge or diverge.

(A) $\sum_{n=0}^{\infty} c_n$

(B) $\sum_{n=0}^{\infty} c_n 6^n$

(C) $\sum_{n=0}^{\infty} c_n (-2)^n$

(D) $\sum_{n=0}^{\infty} (-1)^n c_n 7^n$

500. (A) Express the function $\dfrac{\sin x}{x}$ as a power series.

(B) Evaluate the indefinite integral $\displaystyle\int \dfrac{\sin x}{x}\, dx$ as an infinite series.

(C) Approximate the definite integral $\displaystyle\int_0^1 \dfrac{\sin x}{x}\, dx$ to within an accuracy of three decimal places.

ANSWERS

Chapter 1: Limits and Continuity

1. (C) $\lim_{x \to \pi/4} \tan x = \tan \dfrac{\pi}{4} = 1$

2. (B) $\lim_{t \to -3} \dfrac{t+3}{t^2+9} = \dfrac{(-3)+3}{(-3)^2+9} = \dfrac{0}{18} = 0$

3. (B) Substituting 3 into the expression $\dfrac{x^3 + 2x^2 - 9x - 18}{x - 3}$ results in $\dfrac{0}{0}$. But factoring the numerator gives you $\lim_{x \to 3} \dfrac{(x-3)(x+3)(x+2)}{x-3} = \lim_{x \to 3}(x+3)(x+2) = 6 \cdot 5 = 30$.

4. (D) Substituting 0 into the expression $\dfrac{\sqrt{4-t}-2}{t}$ yields $\dfrac{0}{0}$. So you multiply the expression by a convenient form of 1 and obtain $\lim_{t \to 0} \left(\dfrac{\sqrt{4-t}-2}{t} \right)\left(\dfrac{\sqrt{4-t}+2}{\sqrt{4-t}+2} \right) =$

$\lim_{t \to 0} \dfrac{-t}{t(\sqrt{4-t}+2)} = \lim_{t \to 0} \dfrac{-1}{(\sqrt{4-t}+2)} = \dfrac{-1}{\sqrt{4}+2} = \dfrac{-1}{4}$.

5. (C) Substituting 9 into the expression $\dfrac{9-s}{\sqrt{s}-3}$ gives us $\dfrac{0}{0}$. But multiplying top and bottom by the conjugate of the denominator yields

$$\lim_{s \to 9} \left(\dfrac{9-s}{\sqrt{s}-3} \right)\left(\dfrac{\sqrt{s}+3}{\sqrt{s}+3} \right) = \lim_{s \to 9} \dfrac{(9-s)(\sqrt{s}+3)}{(s-9)} = \lim_{s \to 9} -\sqrt{s} - 3 = -\sqrt{9} - 3 = -6$$

6. (D) The expression $\dfrac{x+4}{x^2+16}$ must be rewritten so that you can apply the Limit Theorem. You do this by multiplying numerator and denominator by $\dfrac{1}{x^2}$.

$$\lim_{x \to \infty} \dfrac{\dfrac{1}{x}+\dfrac{4}{x^2}}{1+\dfrac{16}{x^2}} = \dfrac{\lim_{x \to \infty}\dfrac{1}{x} + \lim_{x \to \infty}\dfrac{4}{x^2}}{\lim_{x \to \infty}1 + \lim_{x \to \infty}\dfrac{16}{x^2}} = \dfrac{0+0}{1+0} = 0$$

7. (A) Since $-1 \le \sin\theta \le 1$ as θ decreases without bound, $\lim_{\theta \to -\infty} \dfrac{\sin\theta}{\theta} = 0$.

8. (B) Multiplying the denominator by $\dfrac{1}{t}$ and the numerator by $\dfrac{1}{\sqrt[3]{t^3}} = \dfrac{1}{t}$ yields

$$\lim_{t \to \infty} \dfrac{\sqrt[3]{1 - \dfrac{8}{t^3}}}{2} = \dfrac{\sqrt[3]{1}}{2} = \dfrac{1}{2}.$$

9. (B) Substituting 0 into the expression $\dfrac{4x^2}{1-\cos 2x}$ results in $\dfrac{0}{0}$. So you multiply the numerator and denominator by the conjugate of $1-\cos 2x$ and get

$$\lim_{x\to 0}\left(\frac{4x^2}{1-\cos 2x}\right)\left(\frac{1+\cos 2x}{1+\cos 2x}\right)=\lim_{x\to 0}\frac{4x^2(1+\cos 2x)}{1-\cos^2 2x}=\lim_{x\to 0}\frac{4x^2}{\sin^2 2x}\cdot\lim_{x\to 0}(1+\cos 2x)$$

$$=\left(\lim_{x\to 0}\frac{2x}{\sin 2x}\right)^2\cdot\lim_{x\to 0}(1+\cos 2x)=1\cdot 2=2$$

10. (B) Substituting 0 into the expression results in $\dfrac{0}{0}$. Factoring the numerator and denominator, however, allows you to simplify.

$$\lim_{\theta\to 0}\frac{(\cos\theta-1)(\cos\theta+1)}{\theta(\cos\theta+1)}=\lim_{\theta\to 0}\frac{(\cos\theta-1)}{\theta}=0$$

11. (D) Again, substituting 0 into the expression gives you $\dfrac{0}{0}$. But remembering trigonometric definitions makes short work of this problem.

$$\lim_{y\to 0}\frac{(\tan y)(\cos y)}{y}=\lim_{y\to 0}\frac{\left(\dfrac{\sin y}{\cos y}\right)(\cos y)}{y}=\lim_{y\to 0}\frac{\sin y}{y}=1.$$

12. (D) Substituting -1 into the expression gets you no nearer a solution, but factoring the numerator will.

$$\lim_{z\to -1}\frac{z^3+1}{z+1}=\lim_{z\to -1}\frac{(z+1)(z^2-z+1)}{z+1}=\lim_{z\to -1}z^2-z+1=3.$$

13. (A) To find the horizontal asymptote(s) you find the limits of the function as $t\to\infty$ and $t\to -\infty$.

$$\lim_{t\to\infty}\frac{27t-18}{3t+8}=\lim_{t\to\infty}\frac{27-\dfrac{18}{t}}{3+\dfrac{8}{t}}=\frac{27}{3}=9$$

Likewise,

$$\lim_{t\to -\infty}\frac{27t-18}{3t+8}=\lim_{t\to -\infty}\frac{27-\dfrac{18}{t}}{3+\dfrac{8}{t}}=\frac{27}{3}=9$$

14. (A) To find the vertical asymptote(s) you must find the values of x for which the expression is undefined, then find left-hand and right-hand limits for those values of x. The expression $\dfrac{x^2+2x+1}{x^2-1}$ is undefined for $x=\pm 1$.

$$\lim_{x\to 1^+}\frac{x^2+2x+1}{x^2-1}=\lim_{x\to 1^+}\frac{(x+1)(x+1)}{(x-1)(x+1)}=\lim_{x\to 1^+}\frac{x+1}{x-1}=\infty$$

$$\lim_{x\to 1^-}\frac{x^2+2x+1}{x^2-1}=\lim_{x\to 1^-}\frac{(x+1)(x+1)}{(x-1)(x+1)}=\lim_{x\to 1^-}\frac{x+1}{x-1}=-\infty$$

Therefore, $x = 1$ is a vertical asymptote of the function. But since $\lim_{x \to -1} \dfrac{x^2 + 2x + 1}{x^2 - 1} =$ $\lim_{x \to -1} \dfrac{x + 1}{x - 1} = 0$, $x = -1$ is not an asymptote.

15. (D) Since $\lim_{x \to 5/2} \dfrac{6x^2 - 11x - 10}{2x - 5} = \lim_{x \to 5/2} \dfrac{(2x - 5)(3x + 2)}{2x - 5} = \lim_{x \to 5/2} 3x + 2 = \dfrac{25}{2}$,

defining $h = \dfrac{25}{2}$ would make the function continuous at $\dfrac{5}{2}$.

16. (E) $g(4) = 15$. So you must find a value of k which $-3(4) - k = 15$. The solution is $k = -27$.

17. (D) By definition, $h(2) = -3$. So you need a value for m such that $(2)^2 - 7(2) + m = -3$. The solution is $m = 7$.

18. (E) Rational functions are continuous everywhere except where the denominator is 0. So you must solve the equation $2x^3 - 8x^2 - 64x = 2x(x - 8)(x + 4) = 0$ to find that the function is discontinuous when $x = 0$, 8 or -4.

19. (E) For $g(x)$ to be continuous, you must have $\lim_{x \to 3} \dfrac{x^2 + 4x - 21}{x^2 - 8x + 15} = 5$ and $\lim_{x \to 5} \dfrac{x^2 + 4x - 21}{x^2 - 8x + 15} = 1$.

$$\lim_{x \to 3} \frac{x^2 + 4x - 21}{x^2 - 8x + 15} = \lim_{x \to 3} \frac{x + 7}{x - 5} = \frac{10}{-2} = -5$$

So $g(x)$ is not continuous at $x = 3$.

$$\lim_{x \to 5} \frac{x^2 + 4x - 21}{x^2 - 8x + 15} = \lim_{x \to 5} \frac{x + 7}{x - 5} = \infty$$

So $g(x)$ is discontinuous at $x = 5$. Only I and III do not contain this point.

20. (C) By the Intermediate Value Theorem, only 6 is a possible value for a.

21. (A) The discontinuities at $x = 1$ and $x = 2$ can be removed by defining $f(1) = -\dfrac{1}{2}$ and $f(2) = -\dfrac{1}{5}$. There is no discontinuity at $x = 3$. However, $\lim_{x \to -3} \dfrac{x^3 - 6x^2 + 11x - 6}{x^3 - 7x + 6} =$ $\lim_{x \to -3} \dfrac{(x - 1)(x - 2)(x - 3)}{(x - 1)(x - 2)(x + 3)} = \lim_{x \to -3} \dfrac{(x - 3)}{(x + 3)}$ does not exist. Therefore, the discontinuity at $x = -3$ is not removable.

22. (D) $\lim_{\theta \to \pi/4} \dfrac{\cos 2\theta}{\cos \theta - \sin \theta} = \lim_{\theta \to \pi/4} \dfrac{\cos^2 \theta - \sin^2 \theta}{\cos \theta - \sin \theta} = \lim_{\theta \to \pi/4} \dfrac{(\cos \theta - \sin \theta)(\cos \theta + \sin \theta)}{\cos \theta - \sin \theta} =$

$\lim_{\theta \to \pi/4} (\cos \theta + \sin \theta) = \sqrt{2}$

23. (B) For $t = a$ to be a removable discontinuity of the function, the numerator and denominator must both evenly divide by $(t - a)$. Top and bottom both divide by $(t - 2)$, but not $(t - 1)$, $(t - 3)$, $(t - 4)$ or $(t - 6)$. The removable discontinuity is $t = 2$.

24. (E) $\dfrac{x^2 + x - 12}{x - 7x + 12} = \dfrac{(x - 3)(x + 4)}{(x - 3)(x - 4)}$, so the discontinuity at $x = 3$ is removable.

Furthermore, $\lim\limits_{x \to 4^+} \dfrac{x^2 + x - 12}{x - 7x + 12} = \lim\limits_{x \to 4^+} = \dfrac{(x - 3)(x + 4)}{(x - 3)(x - 4)} = \lim\limits_{x \to 4^+} \dfrac{x + 4}{x - 4} = \infty$, but

$\lim\limits_{x \to 4^-} \dfrac{x^2 + x - 12}{x - 7x + 12} = \lim\limits_{x \to 4^-} = \dfrac{(x - 3)(x + 4)}{(x - 3)(x - 4)} = \lim\limits_{x \to 4^-} \dfrac{x + 4}{x - 4} = -\infty$. So the vertical asymptote is at $x = 4$.

25. (A) $\lim\limits_{x \to -7/2} \dfrac{6x^2 + 5x - 56}{2x + 7} = \lim\limits_{x \to -7/2} \dfrac{(2x + 7)(3x - 8)}{2x + 7} = \lim\limits_{x \to -7/2}(3x - 8) = -\dfrac{37}{2}$

26. (D) Since $g(5) = 17$, h must be defined so that $2(5)^3 - 9(5)^2 - 2(5) + h = 17$. Solving this equation for h yields $h = 32$.

27. (E) By factoring, you see that $\dfrac{x^3 + 5x^2 - 2x - 24}{x^3 - 3x^2 - 10x + 24} = \dfrac{(x - 2)(x + 3)(x + 4)}{(x - 2)(x + 3)(x - 4)} = \dfrac{(x + 4)}{(x - 4)}$.
So $\lim\limits_{x \to 4} f(x)$ does not exist, and the point of discontinuity $x = 4$ is not removable.

28. (E) The function is discontinuous at $x = \dfrac{3}{2}$. But that point is not included in the intervals given in I and III.

29. (B) To find the horizontal asymptote(s) you find the limits of the function as $x \to \infty$ and $x \to -\infty$.

$$\lim\limits_{x \to \infty} \frac{x}{x^2 + 1} = \lim\limits_{x \to \infty} \frac{\dfrac{1}{x}}{1 + \dfrac{1}{x^2}} = \frac{0}{1} = 0$$

Likewise,

$$\lim\limits_{x \to -\infty} \frac{x}{x^2 + 1} = \lim\limits_{x \to -\infty} \frac{\dfrac{1}{x}}{1 + \dfrac{1}{x^2}} = \frac{0}{1} = 0$$

30. (C) Simplifying the rational expression by factoring gives

$$(t) = \frac{x^4 - 7x^3 + 5x^2 + 31x - 30}{x^4 + x^3 - 19x^2 + 11x + 30} = \frac{(x - 1)(x + 2)(x - 3)(x - 5)}{(x + 1)(x - 2)(x - 3)(x + 5)} = \frac{(x - 1)(x + 2)(x - 5)}{(x + 1)(x - 2)(x + 5)}$$

So the only removable discontinuity is at $x = 3$.

31. (C) $\lim\limits_{x \to 1} \dfrac{x^4 - x^3 + x - 1}{x - 1} = \lim\limits_{x \to 1} \dfrac{(x^3 + 1)(x - 1)}{x - 1} = \lim\limits_{x \to 1} x^3 + 1 = 2$

32. (C) Simplifying the rational expression by factoring gives you

$$\frac{2x^3 + x^2 - 25x + 12}{2x^3 + 3x^2 - 23x - 12} = \frac{(2x - 1)(x + 4)(x - 3)}{(2x + 1)(x + 4)(x - 3)} = \frac{(2x - 1)}{(2x + 1)}$$

So the discontinuity that is not removable is at $x = -\dfrac{1}{2}$.

33. (C) $\lim_{x\to 0}\dfrac{\sqrt{x^2+1}-1}{x}=\lim_{x\to 0}\dfrac{(\sqrt{x^2+1}-1)(\sqrt{x^2+1}+1)}{x(\sqrt{x^2+1}+1)}=\lim_{x\to 0}\dfrac{x^2}{x(\sqrt{x^2+1}+1)}=$

$\lim_{x\to 0}\dfrac{x}{(\sqrt{x^2+1}+1)}=\dfrac{0}{2}=0$

34. (B) $\lim_{t\to 2}\dfrac{t^2-4}{t^3-8}=\lim_{t\to 2}\dfrac{(t-2)(t+2)}{(t-2)(t^2+2t+4)}=\dfrac{1}{3}$

35. (E) Since $\lim_{x\to -1^+}\dfrac{x^2+1}{x+1}\neq\lim_{x\to -1^+}\dfrac{x^2+1}{x+1}$ the limit does not exist.

36. (D) Since the function is continuous and never attains a value of -1 on the interval, neither will it attain any value greater than -1. Therefore, only -4 and -3 are possible values for a.

37. (A) Since $\lim_{x\to -4}\dfrac{x^2-2x-24}{x^2+10x+24}=\lim_{x\to -4}\dfrac{(x-6)(x+4)}{(x+6)(x+4)}=\lim_{x\to -4}\dfrac{(x-6)}{(x+6)}=\dfrac{-10}{2}=$

$-5=g(-4)$, the function is continuous at $x=-4$. But $\lim_{x\to -6}\dfrac{x^2-2x-24}{x^2+10x+24}$ does not exist,

since the left-hand limit is $-\infty$ but the right-hand limit is ∞. So $g(x)$ is discontinuous at $x=-6$. Only the interval given by I omits that point.

38. (B) $\lim_{x\to 0}\dfrac{\sqrt{6-x}-\sqrt{6}}{x}=\lim_{x\to 0}\dfrac{\sqrt{6-x}-\sqrt{6}}{x}\cdot\dfrac{\sqrt{6-x}+\sqrt{6}}{\sqrt{6-x}+\sqrt{6}}=$

$\lim_{x\to 0}\dfrac{-1}{\sqrt{6-x}+\sqrt{6}}=-\dfrac{1}{2\sqrt{6}}=-\dfrac{\sqrt{6}}{12}.$

39. (D) $\lim_{x\to 11}\dfrac{\sqrt{x+5}-4}{x-11}=\lim_{x\to 11}\dfrac{\sqrt{x+5}-4}{x-11}\cdot\dfrac{\sqrt{x+5}+4}{\sqrt{x+5}+4}=\lim_{x\to 11}\dfrac{1}{\sqrt{x+5}+4}=\dfrac{1}{8}$

40. (B) $\lim_{x\to 0}\dfrac{\sin x^3}{x^2}=\lim_{x\to 0}\dfrac{x\sin x^3}{x^3}=\lim_{x\to 0}x\cdot\lim_{x\to 0}\dfrac{\sin x^3}{x^3}=0\cdot 1=0$

41. (E) $\lim_{x\to 3}\dfrac{x^2-x}{x-3}=\infty$ since the numerator is approaching 6 while the denominator is approaching 0 positively.

42. (C) $\lim_{x\to\infty}\dfrac{45x^2+13x-18}{4x-9x^3}=\lim_{x\to\infty}\dfrac{\dfrac{45}{x}+\dfrac{13}{x^2}-\dfrac{18}{x^3}}{\dfrac{4}{x^2}-9}=0$

43. (A) In order for the function to be continuous at $x = -\frac{7}{2}$, $f\left(-\frac{7}{2}\right)$ must equal $\lim_{x \to -7/2} f(x)$. And since

$$\lim_{x \to -7/2} \frac{28x^2 - 13x - 6}{7x + 2} = \lim_{x \to -7/2} \frac{(4x - 3)(7x + 2)}{7x + 2} = \lim_{x \to -7/2} (4x - 3) = \frac{-23}{3}$$

the value of k must also equal $\dfrac{-23}{3}$.

44. (E) In the neighborhood $x = 5$, the numerator is always negative, but the sign of the denominator will change depending upon whether $x < 5$ or $x > 5$. And since

$$\lim_{x \to 5^+} \frac{x^2 - 24}{5 - x} = \infty \neq \lim_{x \to 5^-} \frac{x^2 - 24}{5 - x} = -\infty$$

the limit does not exist.

45. (C) $\lim\limits_{x \to \infty} \dfrac{\sqrt{x} - 7}{6 - 5\sqrt{x}} = \lim\limits_{x \to \infty} \dfrac{1 - \dfrac{7}{\sqrt{x}}}{\dfrac{6}{\sqrt{x}} - 5} = -\dfrac{1}{5}$

46. (A) $\lim\limits_{x \to \pi/4} \dfrac{x(1 - \tan x)}{\cos x - \sin x} = \lim\limits_{x \to \pi/4} \dfrac{x\left(\dfrac{\cos x - \sin x}{\sin x}\right)}{\cos x - \sin x} = \lim\limits_{x \to \pi/4} \dfrac{x}{\sin x} = \dfrac{2\pi}{4\sqrt{2}} = \dfrac{\pi\sqrt{2}}{4}$

(B) Near $x = \dfrac{\pi}{4}$, $x \cdot \sec x = \dfrac{\pi\sqrt{2}}{4}$. This implies that $\sec x = \dfrac{\pi\sqrt{2}}{4x}$.

47. (A) The area of a triangle inscribed as shown is given by $\dfrac{1}{2}bh = \dfrac{1}{2}r^2 \sin\theta$. Since there can be $\dfrac{2\pi}{\theta}$ triangles of the same size inscribed in the circle, the total area of the n-gon is given by $\dfrac{2\pi}{\theta} \cdot \dfrac{r^2 \sin\theta}{2}$. As $\theta \to 0$, the area of this polygon will more closely match the area of the circle.

$$\lim_{\theta \to 0} \frac{\pi r^2 \sin\theta}{\theta} = \pi r^2 \lim_{\theta \to 0} \frac{\sin\theta}{\theta} = \pi r^2$$

(B)

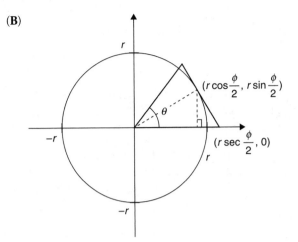

The area of a triangle circumscribed as shown is given by $2\left(\dfrac{1}{2}\right)\left(r\sec\dfrac{\varphi}{2}\right)\left(r\sin\dfrac{\varphi}{2}\right) = r^2\tan\dfrac{\varphi}{2}$. Since there can be $\dfrac{2\pi}{\varphi}$ such triangles circumscribed, the total area of an n-gon composed of such triangles is given by $\dfrac{2\pi}{\varphi}\cdot r^2\tan\dfrac{\varphi}{2}$. As $\varphi\to 0$, the area of this polygon will more closely match the area of the circle.

$$\lim_{\varphi\to 0}\frac{\pi r^2\tan\dfrac{\varphi}{2}}{\varphi}=\pi r^2\cdot\lim_{\varphi\to 0}\frac{\sin\dfrac{\varphi}{2}}{\varphi}\cdot\lim_{\varphi\to 0}\sec\frac{\varphi}{2}=\pi r^2$$

48. (A) Polynomials of even degree do not necessarily have a root, so $f(x)$ need not have any discontinuities. Also, $f(x)$ may have as many as four discontinuities. These continuities will only be removable, where p and q have roots in common.

   (B) Since polynomials of odd degree will have at least one root, $f(x)$ will have at least one discontinuity, and as many as five. But the polynomial need not have any non-removable discontinuities due to multiplicity of roots.

   (C) This situation is the same as the first. Roots of the numerator do not produce discontinuities.

49. $0.\overline{99}=\lim_{x\to\infty}(1-10^{-x})=1-0=1$

50. $\displaystyle\lim_{x\to 0}\frac{4-\sqrt{16-x}}{x}=\lim_{x\to 0}\frac{4-\sqrt{16-x}}{x}\cdot\frac{4+\sqrt{16-x}}{4+\sqrt{16-x}}=\lim_{x\to 0}\frac{x}{x(4+\sqrt{16-x})}$

$$=\lim_{x\to 0}\frac{1}{(4+\sqrt{16-x})}=\frac{1}{4}$$

Chapter 2: Differentiation

51. (B)

$$\lim_{\Delta x \to 0} \frac{\sin\left(\dfrac{\pi}{4} + \Delta x\right)\cos\left(\dfrac{\pi}{4} + \Delta x\right) - \sin\left(\dfrac{\pi}{4}\right)\cos\left(\dfrac{\pi}{4}\right)}{\Delta x} = \lim_{\Delta x \to 0} \frac{\dfrac{1}{2}\sin\left(\dfrac{\pi}{2} + 2\Delta x\right) - \dfrac{1}{2}\sin\left(\dfrac{\pi}{2}\right)}{\Delta x}$$

$$= \frac{1}{2}\lim_{\Delta x \to 0} \frac{\sin\left(\dfrac{\pi}{2}\right) \cdot \cos(2\Delta x) + \cos\left(\dfrac{\pi}{2}\right) \cdot \sin(2\Delta x) - \sin\left(\dfrac{\pi}{2}\right)}{\Delta x}$$

$$= \lim_{2\Delta x \to 0} \frac{\cos(2\Delta x) - 1}{2\Delta x} = 0$$

52. (E)

$$\lim_{h \to 0} \frac{\tan\left(\dfrac{\pi}{4} + h\right) - \tan\left(\dfrac{\pi}{4}\right)}{h} = \lim_{h \to 0} \frac{1}{h} \cdot \left[\frac{\tan\left(\dfrac{\pi}{4}\right) + \tan(h)}{1 - \tan\left(\dfrac{\pi}{4}\right)\tan(h)} - \tan\left(\dfrac{\pi}{4}\right) \right]$$

$$= \lim_{h \to 0}\left[\frac{1}{h} \cdot \frac{1 + \tan(h)}{1 - \tan(h)} - 1 \right] = \lim_{h \to 0}\left[\frac{1}{h} \cdot \frac{1 + \tan(h) - (1 - \tan(h))}{1 - \tan(h)} \right]$$

$$= \lim_{h \to 0}\left[\frac{1}{h} \cdot \frac{2\tan(h)}{1 - \tan(h)} \cdot \frac{\cos(h)}{\cos(h)} \right] = \lim_{h \to 0}\left[\frac{2}{h} \cdot \frac{\sin(h)}{\cos(h) - \sin(h)} \right]$$

$$= 2\left(\lim_{h \to 0} \frac{\sin(h)}{h} \right)\left(\lim_{h \to 0} \frac{1}{\cos(h) - \sin(h)} \right) = 2 \cdot 1 \cdot 1 = 2$$

53. (E) $f'(x) = \cos(x) \cdot \cos^2(\pi - x) + 2\sin(x)\cos(\pi - x)\sin(\pi - x)$

Then $f'\left(\dfrac{\pi}{4}\right) = \dfrac{\sqrt{2}}{2} \cdot \left(-\dfrac{\sqrt{2}}{2}\right)^2 + 2\left(\dfrac{\sqrt{2}}{2}\right)\left(-\dfrac{\sqrt{2}}{2}\right)\left(\dfrac{\sqrt{2}}{2}\right) = \dfrac{\sqrt{2}}{4} + \dfrac{\sqrt{2}}{2} = \dfrac{3\sqrt{2}}{4}$.

54. (C) Implicit differentiation gives us $6x + 3y^2 \cdot y' = 0 \Rightarrow 3y \cdot y' = -6x \Rightarrow y' = \dfrac{-2x}{y^2}$.
Thus, in order to find $y'(3)$, you need to know the value of y when $x = 3$. Plugging 4 in for x
in the original equation results in $27 + y^3 = -37 \Rightarrow y = -4$. So $y'(3) = \dfrac{-2(3)}{(-4)^2} = -\dfrac{6}{16} = -\dfrac{3}{8}$.

55. (E) $x = 0 \Rightarrow y = 0$

Implicit differentiation yields $-\sin x + y \cdot y' = 0 \Rightarrow y \cdot y' = \sin x \Rightarrow y' = \dfrac{\sin x}{y}$.
Hence $y'(0)$ is undefined.

56. (D) Once again, you must use implicit differentiation. Then you obtain

$$\sec y \cdot \tan y \cdot y' = 3(y - x)^2(y' - 1) = (3y - 3x)(y' - 1) = y'(3y - 3x) - (3y - 3x)$$

$$y'(\sec y \cdot \tan y - 3x^2 + 6xy - 3y^2) = -3x^2 + 6xy - 3y^2$$

$$y' = \frac{3x^2 - 6xy + 3y^2}{3x^2 - 6xy + 3y^2 - \sec y \cdot \tan y}$$

57. (D) $\dfrac{dy}{dx} = (x^2 + 2x - 1)(5^{x^2+2x-1})(2x + 2) = (2x^3 + 6x^2 + 2x - 2)(5^{x^2+2x-1})$

58. (A) The limit given is the formal definition for the derivative of $f(x) = \ln(2x)$. It's much easier to just differentiate the function. $\dfrac{dy}{dx}(\ln(2x)) = \dfrac{2}{2x} = \dfrac{1}{x}$.

59. (E) $r = 2\cos\left(\dfrac{\theta}{2}\right) \Rightarrow \dfrac{dr}{d\theta} = -2\sin\left(\dfrac{\theta}{2}\right)\left(\dfrac{1}{2}\right) = -\sin\left(\dfrac{\theta}{2}\right)$

$$x = r\cos\theta \Rightarrow \frac{dx}{d\theta} = -r\sin\theta + \cos\theta\frac{dr}{d\theta} = -r\sin\theta - \cos\theta\sin\left(\frac{\theta}{2}\right)$$

$$y = r\sin\theta \Rightarrow \frac{dy}{d\theta} = r\cos\theta + \sin\theta\frac{dr}{d\theta} = r\cos\theta + \sin\theta\sin\left(\frac{\theta}{2}\right)$$

$$\frac{dy}{dx} = \frac{r\cos\theta + \sin\theta\sin\left(\dfrac{\theta}{2}\right)}{r\cos\theta + \sin\theta\sin\left(\dfrac{\theta}{2}\right)}$$

Substituting $\theta = \pi$ into this formula gives us 0, which is the slope of a tangent line to the curve at this point. Therefore the line normal to this curve will have undefined slope.

60. (B) Using the point-slope form of a linear equation we get $y - 6 = 7(x - 5) \Rightarrow y = 7x - 29$.

61. (A) The Product Rule gives you $h'(2) = f'(2)g(2) + f(2)g'(2) = 5(-1) + 0(3) = -5$.

62. (C) The Chain Rule gives you $h'(3) = g'(f(3)) \cdot f'(3) = g'(1) \cdot (-2) = (-4)(-2) = 8$.

63. (C) You must use implicit differentiation to obtain

$$8x^3 - y - x \cdot \frac{dy}{dx} + 9y^2 \cdot \frac{dy}{dx} = 0 \Rightarrow (9y^2 - x)\frac{dy}{dx} = y - 8x^3 \Rightarrow \frac{dy}{dx} = \frac{y - 8x^3}{9y^2 - x}$$

64. (D)

$$f(x) = 4\sec^3(5x) \Rightarrow f'(x) = (12\sec^2(5x))(\sec(5x)\tan(5x) \cdot 5) = 60\sec^3(5x)\tan(5x)$$

65. (B) $y = \dfrac{e^{3x^2}}{6} \Rightarrow y' = \dfrac{e^{3x^2}}{6} \cdot 6x = x \cdot e^{3x^2} \Rightarrow y'' = e^{3x^2} + 6x^2 \cdot e^{3x^2}$

66. (A) Using the Quotient Rule on $y = \dfrac{2x+7}{5-2x}$ gives you $\dfrac{dy}{dx} = \dfrac{(5-2x)(2)-(2x+7)(-2)}{(5-2x)^2} = $

$\dfrac{10-4x+4x+14}{(5-2x)^2} = \dfrac{24}{(5-2x)^2}.$

67. (B) Applying the Chain Rule to $f(x) = \ln\left(\dfrac{1}{x}\right)$ results in $f'(x) = \dfrac{-x^{-2}}{\dfrac{1}{x}} = -\dfrac{1}{x}.$

68. (C) Since $f(x) = -\cos^2(x^2+2x-3)$ is a composite function you must use the Chain Rule and obtain

$$f'(x) = (-2\cos(x^2+2x-3))(-\sin((x^2+2x-3)))(2x+2)$$
$$= (4x+4)\cos(x^2+2x-3)\sin(x^2+2x-3)$$

69. (E) For $y = \dfrac{e^{x^2}}{x}$ you must use both the Chain Rule and the Quotient Rule.

$$y' = \dfrac{x(2x)e^{x^2} - e^{x^2}}{x^2} = \dfrac{e^{x^2}(2x^2-1)}{x^2}$$

70. (D) $f'(x) = \dfrac{\dfrac{1}{1-x^2}}{\arcsin(x)} \Rightarrow f'\left(\dfrac{\sqrt{2}}{2}\right) = \dfrac{1-\dfrac{1}{2}}{\dfrac{\pi}{4}} = 2 \cdot \dfrac{4}{\pi} = \dfrac{8}{\pi}$

71. (D) The limit given is the definition of the derivative of $f(x) = \sin(x)\cos(x)\ln(x)$ evaluated at $x = \dfrac{\pi}{4}$.

$$f'(x) = \cos^2 x \ln x - \sin^2 x \ln x + \dfrac{\cos x \sin x}{x} \Rightarrow f'\left(\dfrac{\pi}{4}\right) = \dfrac{2}{\pi}$$

72. (A) The limit given is the definition of the derivative of $g(x) = \sec(x)$ evaluated at $x = \dfrac{4\pi}{3}$.

$$g'(x) = \sec x \tan x \Rightarrow g'\left(\dfrac{4\pi}{3}\right) = \sec\dfrac{4\pi}{3}\tan\dfrac{4\pi}{3} = -2\sqrt{3}$$

73. (A) Since $f(x)$ is the product of two functions, you will have to use the product rule. And since $2\sin(2x - \pi)$ and $\cos(3x)$ are composite functions, so you will have to use the chain rule.

$$f'(x) = \cos(3x) \cdot \frac{d}{dx}(\sin^2(2x - \pi)) + \frac{d}{dx}(\cos(3x)) \cdot \sin^2(2x - \pi)$$

$$= \cos(3x) \cdot 4\sin(2x - \pi) \cdot \cos(2x - \pi) + (-3\sin(3x)) \cdot \sin^2(2x - \pi)$$

$$= 4\cos(3x)\sin(2x - \pi) \cdot \cos(2x - \pi) - 3\sin(3x)\sin^2(2x - \pi)$$

$$f'\left(\frac{\pi}{3}\right) = 4\cos(\pi)\sin\left(-\frac{\pi}{3}\right)\cos\left(-\frac{\pi}{3}\right) - 3\sin(\pi)\sin^2\left(-\frac{\pi}{3}\right)$$

$$= 4(-1)\left(-\frac{\sqrt{3}}{2}\right)\left(\frac{1}{2}\right) - 3(0)\left(\frac{3}{4}\right) = \sqrt{3}$$

74. (E) Using the distributive property (to eliminate the parentheses) gives you $y^2x^2 + y^3 = 3x^2$. Now you use implicit differentiation and solve for y'.

$$2xy^2 + x^2 2yy' + 4y^3 y' = 6x$$

$$y'(x^2 2y + 4y^3) + 2xy^2 = 6x$$

$$y' = \frac{6x - 2xy^2}{2x^2 y + 4y^3}$$

Now you need only substitute the values of $x = 2$ and $y = \sqrt{2}$ into this formula to obtain

$$\frac{6(2) - 2(2)(\sqrt{2})^2}{2(2)^2(\sqrt{2}) + 4(\sqrt{2})^3} = \frac{12 - 8}{8\sqrt{2} + 8\sqrt{2}} = \frac{4}{16\sqrt{2}} = \frac{1}{4\sqrt{2}} = \frac{\sqrt{2}}{8}$$

75. (C) You use implicit differentiation to obtain

$$\pi\cos(\pi x) - 9\pi y' \sin(\pi y) = 2xy + x^2 y'$$

$$\pi\cos(\pi x) - 2xy = y'(9\pi\sin(\pi y) - x^2)$$

$$y' = \frac{\pi\cos(\pi x) - 2xy}{9\pi\sin(\pi y) - x^2}$$

When you evaluate this last expression at $(3, -1)$ you obtain $\dfrac{6 - \pi}{9}$.

76. (B) Using implicit differentiation you get

$$2xy^2 + 2x^2 yy' - 3 = 0$$

$$y'(2x^2 y) = 3 - 2xy^2$$

$$y' = \frac{3 - 2xy^2}{2x^2 y}$$

77. **(B)** In general, $[f(g(x))]' = f'(g(x))g'(x)$. Since $f'(x) = \dfrac{1}{4\sqrt{x}}$ and $g'(x) = -\sin x$,

$[f(g(x))]' = \dfrac{1}{4\sqrt{\cos x}}(-\sin x)$.

78. **(D)** At $x = 2$, $\left(\dfrac{f(x)}{g(x)}\right)'$ is $\dfrac{f'(2)g(2) - f(2)g'(2)}{(g(2))^2} = \dfrac{5(3) - 10(7)}{9} = -\dfrac{55}{9}$.

79. **(C)** By applying the product rule several times you get

$$h''(x) = (f'g + fg')' = (f'g)' + (fg')' = f''g + f'g' + f'g' + fg''$$

At $x = 3$, this would be $(1)(2) + (-5)(1) + (-5)(1) + (-1)(4) = -12$.

80. **(A)**

$$\left(-\frac{1}{4}\log_2(5x^2 - 9)\right)' = -\frac{1}{4}\left(\frac{1}{\ln 2(5x^2 - 9)}\right)(5x^2 - 9)' = -\frac{1}{4}\left(\frac{1}{\ln 2(5x^2 - 9)}\right)(10x)$$

$$= -\frac{10x}{4\ln 2(5x^2 - 9)}$$

81. **(C)**

$$\left(\sqrt{x}\log_4\left(\frac{1}{x}\right)\right)' = (\sqrt{x})'\log_4\left(\frac{1}{x}\right) + \sqrt{x}\left(\log_4\left(\frac{1}{x}\right)\right)'$$

$$= \frac{1}{2\sqrt{x}}\log_4\left(\frac{1}{x}\right) + \sqrt{x}\left(\frac{1}{\ln 4\left(\frac{1}{x}\right)}\right)\left(\frac{1}{x}\right)'$$

$$= \frac{1}{2\sqrt{x}}\log_4\left(\frac{1}{x}\right) + \sqrt{x}\left(\frac{1}{\ln 4\left(\frac{1}{x}\right)}\right)\left(-\frac{1}{x^2}\right)$$

$$= \frac{1}{2\sqrt{x}}\log_4\left(\frac{1}{x}\right) - \frac{\sqrt{x}}{x\ln 4} = \frac{\ln 4\log_4\left(\frac{1}{x}\right) - 2}{2\sqrt{x}\ln 4}$$

82. **(E)**

$$\left(\frac{e^{\sqrt{x}}}{x}\right)' = \frac{(e^{\sqrt{x}})'x - e^{\sqrt{x}}(x)'}{x^2} = \frac{e^{\sqrt{x}}\dfrac{1}{2\sqrt{x}}x - e^{\sqrt{x}}}{x^2} = \frac{e^{\sqrt{x}}\dfrac{\sqrt{x}}{2} - e^{\sqrt{x}}}{x^2} = \frac{e^{\sqrt{x}}\left(\dfrac{\sqrt{x}}{2} - 1\right)}{x^2}$$

At $x = 4$,

Evaluating this function at $x = 4$ yields

$$\frac{e^{\sqrt{4}}\left(\dfrac{\sqrt{4}}{2}-1\right)}{4^2} = \frac{e^2\left(\dfrac{2}{2}-1\right)}{16} = \frac{e^2(1-1)}{16} = \frac{e^2(0)}{16} = 0$$

83. **(C)** Note that $\ln(2)$ is a constant.

$$\left(\frac{\ln 2}{2^x}\right)' = \ln 2\left(\frac{1}{2^x}\right)' = \ln 2((2^x)^{-1})' = \ln 2(-1(2^x)^{-2}(2^x)') = \ln 2(-1(2^x)^{-2}2^x \ln 2)$$

$$= -\frac{(\ln 2)^2 2^x}{2^{2x}} = -\frac{(\ln 2)^2}{2^x} = -\frac{\ln^2(2)}{2^x}$$

84. **(C)** Note that e^4 is a constant.

$$e^4(\cos^2(5x))' = e^4(2\cos(5x)(\cos(5x))') = e^4(2\cos(5x)(-\sin(5x))(5x)')$$
$$= -5e^4(2\cos(5x)\sin(5x)) = -5e^4\sin(10x)$$

85. **(D)** By the product rule, $h'(x) = e^x f + e^x f'$. Therefore at $h'(0) = e^0 f(0) + e^0 f'(0) = 2 + 9 = 11$.

86. **(A)** From the problem, you know that $f(3) = 2$ and $f'(3) = 0$. Additionally, by the chain rule $h'(x) = 3(f(x))^2 f'(x)$ and $h'(3) = 3(4)(0) = 0$.

87. **(B)**

$$y' = \cos(x^3)(x^3)' = 3x^2\cos(x^3)$$
$$y'' = (3x^2)'\cos(x^3) + 3x^2(\cos(x^3))' = 6x\cos(x^3) - 3x^2\sin(x^3)(x^3)'$$
$$= 6x\cos(x^3) - 9x^4\sin(x^3) = 3x(2\cos(x^3) - 3x^3\sin(x^3))$$

88. **(D)** When f is differentiable and has an inverse, then that inverse is differentiable at any x, where $f'(f^{-1}(x)) \neq 0$. Further, $(f^{-1})' = \dfrac{1}{f'(f^{-1}(x))}$. Applied to this problem:

$(f^{-1})'(2) = \dfrac{1}{f'(f^{-1}(2))}$. However, in order to use this you must find the value of $f^{-1}(2)$. If $f^{-1}(2) = x$, then $2 = f(x)$, that is $3x^3 + 5 = 2$. So to find $f^{-1}(2)$, you will solve $3x^2 + 5 = 2$. This results in a solution of $x = -1$ which implies that $f(-1) = 2$ and $f^{-1}(2) = -1$. Finally, you must also know $f'(x)$ to complete the process. $f'(x) = 9x^2$. Given this information,

$$(f^{-1})'(2) = \frac{1}{f'(f^{-1}(2))} = \frac{1}{f'(-1)} = \frac{1}{9(-1)^2} = \frac{1}{9}.$$

89. (A) $f(3) = 8$, therefore $3 = f^{-1}(8)$ and $(f^{-1})'(8) = \dfrac{1}{f'(f^{-1}(8))} = \dfrac{1}{f'(3)}$.

90. (C) Evaluating at zero, leads to an indeterminate form: $\lim_{x \to 0} \dfrac{e^{2x} - 1}{\sin x} = \dfrac{e^{2(0)} - 1}{\sin(0)} = \dfrac{1 - 1}{0} = \dfrac{0}{0}$.

Therefore, you can use L'Hopital's rule to evaluate the limit by taking the derivative of the numerator and the denominator.

$$\lim_{x \to 0} \frac{e^{2x} - 1}{\sin x} = \lim_{x \to 0} \frac{(e^{2x} - 1)'}{(\sin x)'} = \lim_{x \to 0} \frac{2e^{2x}}{\cos x} = 2$$

91. (E)

$$\frac{d}{dx}\left(\frac{1}{3}\sqrt{x}\right)^n = \left(\frac{1}{3}\right)^n \frac{d}{dx}(x)^{\frac{n}{2}} = \left(\frac{1}{3}\right)^n \left(\frac{n}{2}\right)(x)^{\frac{n}{2} - 1} = \left(\frac{1}{3}\right)^n \left(\frac{n}{2}\right)(x)^{\frac{n-2}{2}}$$

92. (A) By L'Hopital's rule which states that if a limit results in an indeterminate form, such as 0/0 (which you get here), then you can find the same limit by taking the derivative of both the numerator and the denominator. $\lim_{x \to 0} \dfrac{f}{g} = \dfrac{0}{0}$ since both $f(0)$ and $g(0) = 0$. By L'Hopital,

$$\lim_{x \to 0} \frac{f}{g} = \lim_{x \to 0} \frac{f'}{g'}$$

93. (D) The derivative of any linear function is a constant. Consider the function $y = mx + b$ where both m and b are constants. (Note that m would be a nonzero constant, otherwise this would not be a linear function.) In this case, $y' = m$.

94. (C) A function has a horizontal tangent line if $f' = 0$.
$\dfrac{1}{4}(\sec^2(x))' = \dfrac{1}{4}(2\sec x)(\sec x \tan x) = \dfrac{1}{2}\sec^2 x \tan x$ which is zero at all integer multiples of pi.

95. (C) Since the slopes of the two tangent lines are the same at $x = 2$, the derivatives must be the same at $x = 2$.

96. (A) $f'(x) = \dfrac{2x \sin x \cos x - \sin^2 x}{x^2} = \dfrac{\sin x (2x \cos x - \sin x)}{x^2}$

(B) At $x = \dfrac{\pi}{2}$, $f\left(\dfrac{\pi}{2}\right) = \dfrac{2}{\pi}$, while $f'\left(\dfrac{\pi}{2}\right) = -\dfrac{4}{\pi^2}$. Using the point slope formula:

$$y - \frac{2}{\pi} = -\frac{4}{\pi^2}\left(x - \frac{\pi}{2}\right)$$

$$y = -\frac{4}{\pi^2}x + \frac{4}{\pi}$$

97. (A) $\left(\dfrac{fg}{h}\right)' = \dfrac{(fg)'h - fgh'}{h^2} = \dfrac{(f'g + fg')h - fgh'}{h^2}$

(B) $\dfrac{(e^x \sin x + e^x \cos x)x^2 - e^x \sin x(2x)}{x^4} = \dfrac{xe^x(\sin x + \cos x) - 2\sin x}{x^3}$

$$= \dfrac{e^x((x-2)\sin x) + x\cos x}{x^3}$$

98. $\dfrac{((f(x))^4)'\sqrt[3]{x} - (f(x))^4(\sqrt[3]{x})'}{(\sqrt[3]{x})^2} = \dfrac{4(f(x))^3 f'(x)\sqrt[3]{x} - \dfrac{(f(x))^4}{3x^{2/3}}}{x^{2/3}}$

$$= (f(x))^3\left(\dfrac{4f'(x)}{x^{1/3}} - \dfrac{f(x)}{3x^{4/3}}\right) = (f(x))^3\left(\dfrac{12xf'(x) - f(x)}{3x^{4/3}}\right)$$

99. $f'(x) = \dfrac{2x\cos x + x^2 \sin x}{\cos^2 x}$

$$f\left(\dfrac{\pi}{4}\right) = \dfrac{\dfrac{\pi^2}{16}}{\dfrac{\sqrt{2}}{2}} = \dfrac{\pi^2\sqrt{2}}{16}$$

$$f'\left(\dfrac{\pi}{4}\right) = \dfrac{2\left(\dfrac{\pi}{4}\right)\left(\dfrac{\sqrt{2}}{2}\right) + \dfrac{\pi^2}{16}\left(\dfrac{\sqrt{2}}{2}\right)}{\dfrac{1}{2}} = 2\left(\dfrac{\pi}{2}\left(\dfrac{\sqrt{2}}{2}\right) + \dfrac{\pi^2}{16}\left(\dfrac{\sqrt{2}}{2}\right)\right) = \dfrac{\pi\sqrt{2}(8 + \pi)}{16}$$

Using the point slope formula

$$y - \dfrac{\pi\sqrt{2}(8+\pi)}{16} = \dfrac{\pi^2\sqrt{2}}{16}\left(x - \dfrac{\pi}{4}\right)$$

$$y = \dfrac{\pi^2\sqrt{2}}{16}x - \dfrac{\pi^3\sqrt{2} - \pi^2\sqrt{2}(8+\pi)}{64}$$

$$y = \dfrac{\pi^2\sqrt{2}}{16}x - \dfrac{\pi^2\sqrt{2}(2\pi - 8)}{64}$$

100. While this may appear to involve a difficult derivative, there is no point where both functions have a horizontal tangent line since g is linear and has no horizontal tangent lines.

Chapter 3: Graphs of Functions and Derivatives

101. **(A)** Solve $g'(x) = 0$ for x

$x^2 \sin(x - 1) = 0$

$x^2 = 0$ or $\sin(x - 1) = 0$

$x = 0$ or $x - 1 = -\pi$ or $x - 1 = 0$ ($x - 1 = \pi$ is outside the given interval)

$x = 0$, $x = 1$ or $x = 1 - \pi$

when $x = 0$, 1 or $1 - \pi$ on the interval $-\pi < x \le \pi$.

Since x^2 is always positive, $g'(x)$ is positive when $\sin(x - 1)$ is positive. $g'(x)$ is positive when $x < 1 - \pi$, negative when $1 - \pi < x < 1$, and positive when $x > 1$.

The relative maximum occurs where the derivative is positive to the left and negative to the right at $x = 1 - \pi$.

102. **(B)** Using the product rule $f'(x) = e^{-(x-7)} - xe^{-(x-7)}$

Factor: $e^{-(x-7)} - xe^{-(x-7)} = (1 - x)e^{-(x-7)}$

$f'(x) > 0$ when $1 - x > 0$ since $e^{-(x-7)}$ is always positive.

The function is increasing where $f'(x) > 0$ or when $x < 1$.

103. **(D)** $y' = 2.4x - .4e^{.4x}$

$y'' = 2.4 - .16e^{.4x}$

The function can only change concavity when $y'' = 0$

$0 = 2.4 - .16e^{.4x}$

$.16e^{.4x} = 2.4$

$e^{.4x} = \dfrac{2.4}{.16}$

$e^{.4x} = 15$

$0.4x = \ln 15$

$x = 2.5 \ln 15$

104. **(C)** Using the quotient rule: $f'(x) = \dfrac{1(x - 3) + 1(x)}{(x - 3)^2} = \dfrac{-3}{(x - 3)^2}$.

The slope of the tangent line is when $f'(x) = \dfrac{-3}{4}$.

$\dfrac{-3}{4} = \dfrac{-3}{(x - 3)^2}$

$4 = (x - 3)^2$

$x - 3 = \pm 2$

$x = 3 \pm 2$

$x = 1$ and 5

105. (B) Notice that the derivative is given and you must identify the function.

Where $h'(x)$ passes through zero going from negative to positive, $h(x)$ will have a local minimum. This happens twice.

Where $h'(x)$ passes through zero going from positive to negative, $h(x)$ will have a local minimum. This happens once between the two local minimums.

$h'(x)$ is constant and positive at the end which will correspond to a constant positive slope in $h(x)$.

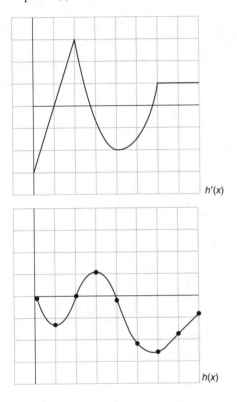

$h'(x)$

$h(x)$

106. (C) The mean value theorem says that $f'(c) = $ the slope of the secant on the given interval.

$$f(1) = 7$$
$$f(b) = 8b - b^3$$

The end points of the secant line are $(1, 7)$ and $(b, f(b))$

The slope of the secant $= \dfrac{f(b) - 7}{b - 1} = \dfrac{8b - b^3 - 7}{b - 1}$

Solve $\dfrac{8b - b^3 - 7}{b - 1} = -\dfrac{1}{2}$ solve for b using the calculator.

One method is to find the zeros of $f(x) = \dfrac{8x - x^3 - 7}{x - 1} + \dfrac{1}{2}$

$B = -3.28$ and $b = 2.28$

Choose $b > 1$ so that $1 < x < b$

$b = 2.28$

107. (D) $p'(x) = 4x^3 - 6x^2 + 10$

$p''(x) = 12x^2 - 12$

An inflection point will occur when $p''(x) = 0$

$12x^2 - 12 = 0$

$12x(x - 1) = 0$, when $x = 0$ or $x = 1$

Inflection points are $p(0) = -8$ and $p(1) = 1$.

Choose $(1, 1)$

108. (A) $f(x) = (9 - x^3)^{1/2}$

Using the chain rule $f'(x) = \dfrac{1}{2}(9 - x^3)^{-1/2} \cdot -3x^2 \dfrac{-3x^2}{2\sqrt{9 - x^3}}$

$f'(2) = \dfrac{-12}{2\sqrt{1}} = -6$

109. (E) The slope of the tangent line at $x = 0$ is $g'(0)$.

Using the chain rule, $g'(x) = \dfrac{2x + 1}{x^2 + x + 6}$

$m = g'(0) = \dfrac{1}{6}$

A point on the tangent line is $g(0) = \ln 6$ or $(0, \ln 6)$

Using the point slope form of a line, $y - \ln 6 = \dfrac{1}{6}(x - 0)$

$y = \dfrac{1}{6}x + \ln 6$

110. (E) The derivative of $g(x)$ should be zero and decreasing at the smooth relative maximum of $g(x)$.

The derivative of $g(x)$ should not exist at the pointed minimum and should jump from being negative on the left to being positive on the right.

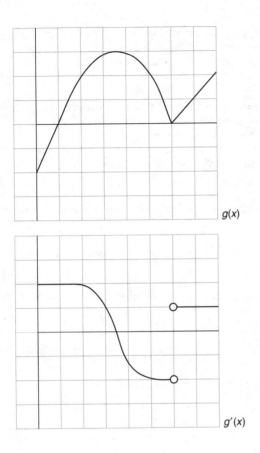

g(x)

g'(x)

111. (A) $\dfrac{dy}{dt} = 2t + 1$, $\dfrac{dx}{dt} = -1$, $\dfrac{dy}{dx} = -2t - 1$, $\dfrac{dy}{dx} > 0$ when $t < -1/2$

112. (A) Using the chain rule, $u = \ln u$, $\dfrac{dy}{du} = \dfrac{1}{u}$, $\dfrac{dy}{dx} = \dfrac{dy}{du}\dfrac{du}{dx} = \dfrac{1}{8 - x^3} \cdot -3x^2 = \dfrac{3x^2}{x^3 - 8}$

113. (C) At $\dfrac{\pi}{4}$, $\dfrac{dx}{dt} = -2\sin(2t) = -2\sin\left(\dfrac{\pi}{2}\right) = -2$, and $\dfrac{dy}{dt} = \cos t = \cos\left(\dfrac{\pi}{4}\right) = \dfrac{\sqrt{2}}{2}$

Thus, $\dfrac{dy}{dx} = \dfrac{\sqrt{2}/2}{-2} = \dfrac{-\sqrt{2}}{4}$.

114. (C) Rolle's theorem requires two points whose secant line has a slope of zero so that $f(a) = f(b)$. (1, 5) and (7, 5) are the only two points which qualify. The interval is $a < x < b$ or $1 < x < 7$.

115. (B) Notice that the information given is about $g(x)$, not $g'(x)$. A function can only change from increasing to decreasing or from decreasing to increasing at a critical point. $g(x)$ increases from −0.33 to 0 on the interval $0 < x < 9.08$ and from −7.3 to −4.4 on the interval $14 < x < 20$. $g(x)$ is increasing on $0 < x < 9.08$ and $14 < x < 20$.

116. (E) The interval $x > 10$ has a positive derivative and $h(x)$ is therefore increasing when $x > 10$. It is not known whether or not the function increases beyond the relative maximum of $h(−2.2)$. $h(x)$ may or may not have a maximum value.

117. (D) A function is concave up where the second derivative is positive. $f''(x) > 0$, when $x < a$ and $x > c$.

118. (D) The slope between $(−3, 8)$ and $(−2, 5) = −3$ which is the slope of the tangent line at $(−3, 8)$ so $f'(−3) = −3$

119. (B) $f(x)$ has an inflection point at $x = b$ since the first derivative goes from increasing to decreasing at $x = b$.

120. (E) Since $g(x)$ is concave up everywhere (except at $x = m$, where $g'(x)$ does not exist), $g''(x)$ will never be negative.

121. (A) $f'(x) = 3x^2 - 1$. The mean value theorem is satisfied when $f'(c) = \dfrac{f(b) - f(a)}{b - a}$.

$$f'(c) = \frac{1^3 - 1 - [(-1)^3 - (-1)]}{1 - (-1)} = \frac{1 - 1 - [(-1) + 1]}{2} = \frac{0 - (-1 - (-1))}{1 - (-1)} = \frac{0}{2} = 0. \ f'(c) =$$

$$3c^2 - 1 = 0 \Rightarrow c = \pm\sqrt{\frac{1}{3}}$$

122. (C) $\cos\theta = \dfrac{r_1 \cdot r_2}{\| r_1 \| \cdot \| r_2 \|} = \dfrac{2(6) + 3(4)}{\sqrt{13} \cdot \sqrt{52}} = \dfrac{24}{26} = \dfrac{12}{13}. \ \theta = \cos^{-1}\left(\dfrac{12}{13}\right) \approx 0.395$ radians.

123. (D) Find the first derivative $f'(x) = x^2 - 12x + 35 = (x - 7)(x - 5)$. The first derivative $f'(x) = 0$, when $x = 5, 7$. Use a test point to figure out if $f'(x) > 0$ or $f'(x) < 0$ for each critical point.

Interval	$(-\infty, 5)$	$(5, 7)$	$(7, +\infty)$
Test points	0	6	10
f'(x)	+	−	+
f(x)	Increasing	Decreasing	Increasing

For the test point $x = 0$, $f'(0) = 0 - 0 + 35 = 35$. For the test point $x = 6$, $f'(6) = 6^2 - 12(6) + 35 = -1$. And for test point $x = 10$, $f'(10) = 10^2 - 12(10) + 35 = 15$. At $x = 5$, $f'(x)$ changes from positive to negative. Therefore, it is a relative maximum. At $x = 7$, $f'(x)$ changes from negative to positive and thus is a relative minimum.

124. (B) Find $f'(x) = 2x^2 + 10x - 28 = 2(x^2 + 5x - 14) = 2(x+7)(x-2)$. $f'(c) = 0$ at $c = -7, 2$.

125. (B) Use the second derivative to find points of inflection. $f'(x) = 6x(x+4)^2 + 2(x+4)^3$. Then we find $f''(x) = 12(x+4) + 6(x+4)^2 + 6(x+4)^2 = 12x(x+4) + 12(x+4)^2$. You can see that $f''(x) = 0$ when $x = -4$. Substituting x in the original function, you find that $f(-4) = 2(-4)(-4+4)^3 = 0$, so $(-4, 0)$ is an inflection point.

126. (D) $x = r\cos\theta$, $y = r\sin\theta$, where $r = 1$. The equation can be rewritten $\dfrac{(r\cos\theta)^2}{25} + \dfrac{(r\sin\theta)^2}{16} = 1$. Multiplying through by 16 and 25, you get $16r^2\cos^2\theta + 25r^2\sin^2\theta = 400$. Isolating r, we get $r^2 = \dfrac{400}{16\cos^2\theta + 25\sin^2\theta}$. Apply the identity $\cos^2\theta = 1 - \sin^2\theta$. Now $r = \dfrac{20}{\sqrt{16 + 9\sin^2\theta}}$.

127. (E) $f'(x) = \cos x$. $f'(c) = \cos c = \dfrac{f\left(\dfrac{\pi}{2}\right) - f(0)}{\dfrac{\pi}{2} - 0} = \dfrac{1-0}{\dfrac{\pi}{2} - 0} = \dfrac{2}{\pi}$. Then $c = \cos^{-1}\left(\dfrac{2}{\pi}\right) \approx 0.8807$.

128. (C) $f'(x) = \dfrac{2}{3}x^{-1/3}$ so $f'(0)$ is not defined. $f(0)$ is, however, defined so there is a critical point at $(0, 0)$. $f''(0)$ is undefined since $f'(0)$ is undefined. $f''(x) = \dfrac{2}{3}\left(-\dfrac{1}{3}\right)x^{-4/3} = -\dfrac{2}{9}x^{-4/3}$. $f''(x) < 0$ when $x \neq 0$ so curve is concave down over intervals $(-\infty, 0)$ and $(0, +\infty)$.

129. (D) A critical point exists when $f'(x) = 0$. Set $f'(x) = 2x^2 - 5 = 0 \Rightarrow 2x^2 = 5 \Rightarrow x = \pm\sqrt{\dfrac{5}{2}}$. Use test points to determine where $f(x)$ is increasing or decreasing.

Interval	$\left(-\infty, -\sqrt{\dfrac{5}{2}}\right)$	$\left(-\sqrt{\dfrac{5}{2}}, \sqrt{\dfrac{5}{2}}\right)$	$\left(\sqrt{\dfrac{5}{2}}, +\infty\right)$
Test point	-2	0	2
$f'(x)$	$+$	$-$	$+$
$f(x)$	Increasing	Decreasing	Increasing

The first derivative $f'(x) < 0$ on interval $\left(-\sqrt{\dfrac{5}{2}}, \sqrt{\dfrac{5}{2}}\right)$ so f is decreasing on $\left(-\sqrt{\dfrac{5}{2}}, \sqrt{\dfrac{5}{2}}\right)$.

130. (A) $x = r\cos\theta = 6\cos\left(\dfrac{\pi}{6}\right) = 6\left(\dfrac{\sqrt{3}}{2}\right) = 3\sqrt{3}$. $y = r\sin\theta = 6\sin\left(\dfrac{\pi}{6}\right) = 6\left(\dfrac{1}{2}\right) = 3$. Therefore the vector is $3\sqrt{3}, 3$.

131. (B) Test for concavity by finding the second derivative. First find $f'(x) = -\dfrac{1}{(x-1)^2}$.
Then find $f''(x) = \dfrac{2}{(x-1)^3}$. There is a discontinuity at $x = 1$ where the denominator equals
zero so evaluate concavity at points where $x > 1$ and $x < 1$. The second derivative $f''(x) < 1$ for
$x < -1$, meaning the function is concave down on $(-\infty, 1)$; $f''(x) > 0$ when $x > 1$, meaning
the function is concave up on $(1, +\infty)$.

132. (E) Since $6\cos(3(-\theta)) = 6\cos(3\theta)$, the graph is symmetric about the x-axis. Then check
for symmetry around the y-axis, or pole. Take $6\cos 3(\theta + \pi) = 6\cos(3\theta + 3\pi)$. This result is
$\neq 6\cos(3\theta)$ so the function is not symmetric about the y-axis. To check for symmetry
around the line $\theta = \dfrac{\pi}{2}$, find $6\cos(3(\pi - \theta)) = 6\cos(3\pi - 3\theta)$ and test if it is equal to
$6\cos 3\theta$. Apply the trigonometric identity $\cos(3\pi - 3\theta) = \cos(3\pi)\cos(3\theta) + \sin(3\pi)\sin(3\theta)$
to get $6[\cos(3\pi)\cos(3\theta) + \sin(3\pi)\sin(3\theta)] = 6[-\cos(3\theta) + 0] = -6\cos(3\theta)$ since $\sin(3\pi) = 0$
and $\cos(3\pi) = -1$. The result $-6\cos(3\theta) \neq 6\cos(3\theta)$. Therefore, it is not symmetric about
the line $\theta = \dfrac{\pi}{2}$.

133. (C) $f'(x) = 3x^2 - 2x$. $f'(x) = 0$, when $3x^2 = 2x$ or $x = 0$, $\dfrac{2}{3}$. $f'(x) > 0$, when $x > \dfrac{2}{3}$
so f is increasing on interval $\left(\dfrac{2}{3}, +\infty\right)$. $f'(x) < 0$ on interval $\left(0, \dfrac{2}{3}\right)$ so f is decreasing on
that interval. $f'(x) > 0$, when $x < 0$ so f is increasing on interval $(-\infty, 0)$.

134. (E) f is a polynomial and thus is continuous on the interval $(-\infty, +\infty)$, so statement I
is true. To test for points of inflection, use the chain rule to find $f'(x) = \dfrac{dy}{du} \cdot \dfrac{du}{dx}$, where $u = $
$x^2 - 4$ and $y = (u^2)^{\frac{1}{2}}$. You then find that $f'(x) = \left(\dfrac{1}{2}\right) \dfrac{2u}{(u^2)^{\frac{1}{2}}} \cdot 2x = \dfrac{2x(x^2 - 4)}{|x^2 - 4|} = \dfrac{2x^3 - 8}{|x^2 - 4|}$.
The first derivative $f'(x)$ has a discontinuity at $x = \pm 2$ since the denominator equals zero at
those values. $f'(x)$I is not differentiable at $x = \pm 2$ so does not have an inflection point there.
Thus statement II is false. The function $f'(x) = 0$ at $x = 0$ and $(0) = 4$, thus has a relative extre-
mum at coordinate $(0, 4)$. Use the second derivative test to see if it is a maximum or minimum.
$f''(x) = \dfrac{dy}{du} \cdot \dfrac{du}{dx} = \dfrac{-2x(2x^3 - 8x)}{|x^2 - 4|^2} + \dfrac{6x^2 - 8}{|x^2 - 4|} = \dfrac{-4x^2(x^2 - 4)}{|x^2 - 4|^2} + \dfrac{6x^2 - 8}{|x^2 - 4|} = \dfrac{2x^2 - 8}{|x^2 - 4|}$. At
$x = 0$, $f''(0) = -2 < 0$ so there is a relative maximum at $x = 0$, $f(0) = 4$. Thus statement III
is true.

135. (C) The magnitude of the vector $r = \langle x, y \rangle$ can be found as $\|r\| = \sqrt{x^2 + y^2} = $
$\sqrt{3^2 + (3\sqrt{3})^2} = \sqrt{9 + 27} = \sqrt{36} = 6$. The angle of rotation is given by $\theta = \tan^{-1}\left(\dfrac{y}{x}\right) = $
$\tan^{-1}\left(\dfrac{3\sqrt{3}}{3}\right) = \tan^{-1}\sqrt{3} = \dfrac{\pi}{3}$.

136. (D) $f'(x) = 8x - 3$. $f'(c)$ is defined for all values of c. $f'(c) = 0$ when $8c = 3$ or $c = \dfrac{3}{8}$.

137. (D) Solve for a and b when $f'(x) = 0$. Set $f'(x) = 3ax^2 - bx^{-2} = 0 \Rightarrow 3ax^2 = \dfrac{b}{x^2} \Rightarrow x^4 = \dfrac{b}{3a}$. You can also substitute the coordinates for the maximum and minimum in the original function to get two expression for a and b. You find that $f(1) = 4 = a + b$ and $f(-1) = -4 = -a - b$. Thus, $a = 4 - b$. You can then use that expression, substituted into $f'(1)$ to get $1 = \dfrac{3a}{b} \Rightarrow 3a = b$. $3(4 - b) = b \Rightarrow b = 3$. Then $a = \dfrac{b}{3} = 1$ or use $a = 4 - b = 1$. You can check our work by plugging the values for a and b into $f'(x)$ and $f''(x)$. If $f(x) = x^3 + \dfrac{3}{x}$, then $f'(x) = 3x^2 - 3x^{-2} = 0$. Then $3x^2 = \dfrac{3}{x^2} \Rightarrow x^4 = \pm 1$. $f''(x) = 6x + \dfrac{6}{x^3} > 0$ for $x > 0$ and $f''(x) < 0$ for $x < 0$.

138. (E) Apply the equation $r = \sqrt{x^2 + y^2}$ to find $r^2 = x^2 + y^2 = \dfrac{1}{\cos 2\theta}$. Apply the trigonometric identity $\cos 2\theta = 1 - 2\sin^2 \theta$ to get $x^2 + y^2 = \dfrac{1}{1 - 2\sin^2 \theta}$. Then apply the relation $y = r\sin\theta$ to find that $\sin^2 \theta = \dfrac{y^2}{r^2} = \dfrac{y^2}{x^2 + y^2}$. Now substitute for $\sin^2 \theta$ to find that

$$x^2 + y^2 = \dfrac{1}{1 - \dfrac{2y^2}{x^2 + y^2}} = \dfrac{1}{\dfrac{x^2 - y^2}{x^2 + y^2}} = \dfrac{x^2 + y^2}{x^2 - y^2} \Rightarrow (x^2 + y^2)(x^2 - y^2) = x^2 + y^2 \text{ or } x^2 - y^2 = 1.$$

139. (E) By inspection you see that f has no discontinuities on $[0, 4]$. Also, $f'(x) = 6x - 12$ and is also continuous on $[0, 4]$. Therefore $f(x)$ is differentiable on the interval $[0, 4]$. Rolle's theorem also requires that $f(a) = f(b) = 0$. In this case $f(0) = f(4) = 1$. This function does not satisfy Rolle's theorem.

140. (C) $f'(x) = 0$ when $x^4 = 16$, or $x = \pm 2$. $f'(x) > 0$ on the intervals $(-\infty, -2)$ and $(2, +\infty)$ therefore $f(x)$ is increasing on those intervals. On the interval $(-2, 2)$, $f'(x) < 0$, therefore f is decreasing.

141. (D) Solve the first equation for t in terms of x, $t = e^x \Rightarrow y = 4e^x + 1$.

142. (D) $f'(x) = 6x^2 + 2x = 0 \Rightarrow 6x^2 = -2x \Rightarrow x = -\dfrac{1}{3}$. The first derivative $f'(x)$ also equals zero when $x = 0$. The second derivative $f''(x) = 12x + 2$. At $x = -\dfrac{1}{3}$, $f''\left(-\dfrac{1}{3}\right) = 12\left(-\dfrac{1}{3}\right) + 2 = -2$ which is < 0 so $f\left(-\dfrac{1}{3}\right)$ is a relative maximum. At $x = 0$, $f''(0) = 2 > 0$ so $f(0)$ is a relative minimum. To find the maximum and minimum points on f, plug values into original function. $f(0) = 15$ and $f\left(-\dfrac{1}{3}\right) = \dfrac{406}{27}$.

143. (B) Use the second derivative to determine concavity. $f''(x) = 6x - 2 = 0$ when $6x = 2$ or $x = \frac{1}{3}$. $f'' < 0$ when $x < \frac{1}{3}$ so the graph of f is concave up on the interval $\left(-\infty, \frac{1}{3}\right)$. $f'' > 0$ for $x > 0$ so f is concave down on the interval $\left(\frac{1}{3}, +\infty\right)$.

144. (C) An equation of the form $r = a\cos(n\theta)$ describes a rose of petal length a, with $2n$ petals since n is an even number; thus $r = 5\cos 4\theta$ yields a rose with $2(4) = 8$ petals of length 5.

145. (E) Statement I is correct since f is strictly increasing on the interval from D to F. Statement II is true since there is a horizontal tangent to the curve at points B, D, and F, implying $f' = 0$. Statement III is true since f is concave downward on the interval from A to B. Statement IV is false since f is concave upward only on the interval from C to D. The function is concave downward from B to C, so $f'' < 0$ for that interval.

146. (A) Use the first derivative to find where f is increasing or decreasing. The first derivative $f'(x) = -\frac{1}{3}x^{-2/3} < 0$ for $(-\infty, +\infty)$, then f is decreasing for all values of x.

(B) First, find any points where $f'(x) = 0$, but $f' < 0$ for all values of x. At $x = 0$, $f'(x)$ does not exist since there is a zero in the denominator. Therefore, there are no maxima or minima.

(C) Use the second derivative to test for concavity. $f''(x) = \frac{2}{9}x^{-5/3}$. $f'' > 0$ when $x > 0$, therefore f is concave upward on the interval $(0, +\infty)$. If $x < 0$, then $f'' < 0$, therefore f is concave downward on $(-\infty, 0)$.

(D) The second derivative f'' changes concavity at $x = 0$, so there is an inflection point at $x = 0$. The function $f(0) = 1$, so the inflection point is at coordinate $(0, 1)$ which is also the y-intercept.

(E) y-intercept at $(0, 1)$. x-intercept at $(1, 0)$.

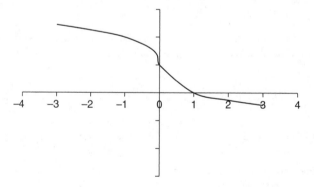

147. (A) If vectors are orthogonal, the angle between them is $\frac{\pi}{2}$ so that their dot product is defined by $r_1 r_2 \cos\theta = 0$. Then $r_1 \cdot r_2 = x_1 x_2 + y_1 y_2 = \langle 1, -4\rangle \cdot \langle 2, k\rangle = 1(2) + (-4)(k) = 0$. Then $-4k = -2 \Rightarrow k = \frac{1}{2}$.

(B) If vectors are parallel, the angle between them is 0, where $\cos(0) = 1$. You know that, in the case of parallel vectors, $r_2 = Cr_1$, where C is a constant. In this case, the horizontal component of r_2 differs from r_1 by a factor of 2. Multiplying the y-component by the same factor, you get $k = -8$. You can get the same result by applying the equation $\cos\theta = \dfrac{r_1 \cdot r_2}{\|r_1\|\|r_2\|} = -1$. You find that $\dfrac{2-4k}{(\sqrt{17})\sqrt{4+k^2}} = -1$. If you multiply through by the denominator and then square both sides, you find that $4 - 16k + 16k^2 = 17(4 + k^2)$ and, combining terms, you get $k^2 + 16k + 64 = 0$. You can then factor to find that $(k + 8)(k + 8) = 0$ or $k = -8$.

(C) Find the angle between two vectors by applying the equation $\theta = \cos^{-1}\dfrac{r_1 \cdot r_2}{\|r_1\|\|r_2\|}$.

Here, $r_1 \cdot r_2 = 1(2) - 4(6) = -22$, $\|r_1\| = \sqrt{1^2 + 4^2} = \sqrt{17}$ and $\|r_2\| = \sqrt{2^2 + 6^2} = \sqrt{40} = 2\sqrt{10}$.

Substituting, we find that $\theta = \dfrac{-22}{2\sqrt{170}} \approx 147.5°$.

148. (A) Draw a vector diagram.

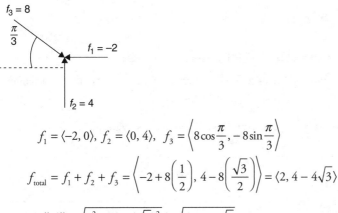

$$f_1 = \langle -2, 0\rangle, \quad f_2 = \langle 0, 4\rangle, \quad f_3 = \left\langle 8\cos\frac{\pi}{3}, -8\sin\frac{\pi}{3}\right\rangle$$

$$f_{total} = f_1 + f_2 + f_3 = \left\langle -2 + 8\left(\frac{1}{2}\right), 4 - 8\left(\frac{\sqrt{3}}{2}\right)\right\rangle = \langle 2, 4 - 4\sqrt{3}\rangle$$

(B) $\|f\| = \sqrt{2^2 + (4 - 4\sqrt{3})^2} = \sqrt{68 + 32\sqrt{3}} \approx 11.11$ newtons.

(C) Apply an equal and opposite force to keep the object from moving so $f_{opp} = \langle -2, -4 + 4\sqrt{3}\rangle$.

149. (A) $f' < 0$ on $(-\infty, -1)$ and $(1, 5)$ so f is decreasing. $f' > 0$ on $(-1, 1)$ and $(5, +\infty)$ so f is increasing.

(B) The function has a relative maximum at $x = 1$ since f' changes from positive to negative. There are relative minima at $x = -1, 5$ since f' changes from negative to positive.

(C) f' is increasing on $(-\infty, 0)$ so $f'' > 0$ and f is concave upward. f' is decreasing on $\left(0, 2\frac{1}{2}\right)$ so $f'' < 0$ and f is concave downward. f' is increasing on $\left(2\frac{1}{2}, +\infty\right)$ so $f'' > 0$ and f is concave upward. A change of concavity occurs at $x = 0$ and $x = 2\frac{1}{2}$, implying the existence of a tangent line at $x = 0$ and $x = 2\frac{1}{2}$. Summarizing the results in a table,

Interval	$(-\infty, 0)$	$\left(0, 2\frac{1}{2}\right)$	$\left(2\frac{1}{2}, \infty\right)$
f'	Increasing	Decreasing	Increasing
f''	+	−	+
f	Concave Upward	Concave Downward	Concave Upward

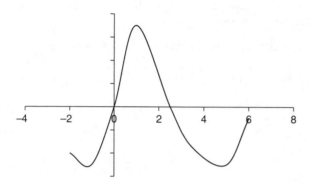

150. (A) To figure out where $f(x)$ is increasing or decreasing, set $f'(x) = 4x^3 - 3x^2 = x^2(4x - 3) = 0$. The function equals zero at $x = 0$ and $x = \dfrac{3}{4}$. Use a test point on each interval to find where $f(x)$ is increasing or decreasing. The function $f(x)$ is increasing on interval $\left(\dfrac{3}{4}, +\infty\right)$, but is decreasing for interval $(-\infty, 0)$ and decreasing for $\left(0, \dfrac{3}{4}\right)$.

Interval	$(-\infty, 0)$	$\left(0, \dfrac{3}{4}\right)$	$\left(\dfrac{3}{4}, +\infty\right)$
Test point	-1	$\dfrac{1}{2}$	1
$f'(x)$	−	−	+
$f(x)$	Decreasing	Decreasing	Increasing

(B) You found in part A that $f'(x) = 0$ at $x = 0$ and $x = \dfrac{3}{4}$. You then take the second derivative at those points to determine if they represent a maximum or minimum. You find that $f''(x) = 12x^2 - 6x$. You then substitute values for x to find that $f''(0) = 0$ and $f''\left(\dfrac{3}{4}\right) = \dfrac{9}{4}$. This demonstrates that $f\left(\dfrac{3}{4}\right)$ is a relative minimum. The test for $f''(0)$ is inconclusive so look back to the first derivative. The first derivative changes from positive to negative at $x = 0$, which implies that $x = 0$ is a relative maximum.

At $x = \dfrac{3}{4}$, $f' = 0$. $f''\left(\dfrac{3}{4}\right) = 12\left(\dfrac{9}{16}\right) - 6\left(\dfrac{3}{4}\right) = \dfrac{108}{16} - \dfrac{72}{16} = \dfrac{36}{16} > 0$ so $x = \dfrac{3}{4}$ is a rela-

tive minimum. $f\left(\dfrac{3}{4}\right) = \left(\dfrac{3}{4}\right)^4 - \left(\dfrac{3}{4}\right)^3 = \dfrac{81}{256} - \dfrac{108}{256} = -\dfrac{27}{256}$ so relative minimum at

$f\left(\dfrac{3}{4}\right) = -\dfrac{27}{256}$.

(C) $f''(x) = 12x^2 - 6x = 6x(2x - 1) = 0$ at $x = 0, \dfrac{1}{2}$. $f'' > 0$ on interval $\left(\dfrac{1}{2}, +\infty\right)$.

$f'' < 0$ on interval $\left(0, \dfrac{1}{2}\right)$. $f'' > 0$ on interval $(-\infty, 0)$. At $x = 0, f = 0$, so $f(0) = 0$ is an inflection

point. At $x = \dfrac{1}{2}$, $f = \left(\dfrac{1}{2}\right)^4 - \left(\dfrac{1}{2}\right)^3 = \dfrac{1}{16} - \dfrac{1}{8} = -\dfrac{1}{16}$ so $f\left(\dfrac{1}{2}\right) = -\dfrac{1}{16}$ is another inflection

point.

(D) Use the second derivative to test for concavity. On $(-\infty, 0)$, $f'' > 0$ so f is con-

cave upward. On $\left(0, \dfrac{1}{2}\right)$, $f'' < 0$ so f is concave downward. On $\left(\dfrac{1}{2}, +\infty\right)$, $f'' > 0$ so f is

concave upward.

Interval	$(-\infty, 0)$	$\left(0, \dfrac{1}{2}\right)$	$\left(\dfrac{1}{2}, +\infty\right)$
f''	+	−	+
Concavity	Up	Down	Up

Chapter 4: Applications of Derivatives

151. (C) You are given a position function: $f(x) = 2x + 5$ in meters. The velocity function is the first derivative of $f(x)$ and its unit is meters per second.

$$f(x) = 2x + 5$$
$$f'(x) = 2$$

So, velocity = 2 m/s

152. (D) Let V be the volume of the sphere, s be the surface area of the sphere, and r be the radius of the sphere. You are given that $dS/dt = 4\ dr/dt$.

$$S = 4\pi r^2$$

$$\frac{dS}{dt} = 8\pi r \frac{dr}{dt}$$

$$\frac{dS}{dt} = 4\frac{dr}{dt} \qquad \text{(Given)}$$

$$8\pi r \frac{dr}{dt} = 4\frac{dr}{dt}$$

$$2\pi r = 1$$

$$r = \frac{1}{2\pi}$$

$$V = \frac{4}{3}\pi r^3$$

$$V = \frac{4}{3}\pi \left(\frac{1}{2\pi}\right)^3$$

$$V = \frac{4}{3}\pi \left(\frac{1}{8\pi^3}\right)$$

$$V = \frac{1}{6\pi^2} \text{ cubic units}$$

153. (D) The base of the funnel is 4 cm, so the radius(r) is 2.0 cm. The height of the funnel is 5 cm. The rate of the drainage (dV/dt) is a constant 2.0 cm/s. So, what is the change in height of the water level (dh/dt) when the height (h) = 2.5 cm?

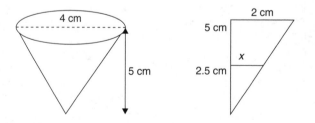

$$\frac{2}{5} = \frac{r}{h} \text{ (Similar triangles)}$$

$$2h = 5r$$

$$r = \frac{2h}{5}$$

$$V = \frac{1}{3}\pi r^2 h$$

$$V = \frac{1}{3}\pi \left(\frac{2h}{5}\right)^2 h$$

$$V = \frac{1}{3}\pi \left(\frac{4h^2}{25}\right) h$$

$$V = \frac{4}{75}\pi h^3$$

$$\frac{dV}{dt} = 3\left(\frac{4}{75}\right)\pi h^2 \frac{dh}{dt}$$

$$\frac{dV}{dt} = \left(\frac{4}{25}\right)\pi h^2 \frac{dh}{dt}$$

$$\frac{dV}{dt} = -2 \text{ cm}^3/s$$

$$-2 = \left(\frac{4}{25}\right)\pi h^2 \frac{dh}{dt}$$

$$-1 = \left(\frac{2}{25}\right)\pi h^2 \frac{dh}{dt}$$

$$\frac{dh}{dt} = -\frac{25}{2\pi h^2}$$

$$\frac{dh}{dt} = -\frac{25}{2\pi(2.5)^2}$$

$$\frac{dh}{dt} = -\frac{2}{\pi} \text{ cm/s}$$

154. (C) Let S be the height of the shadow, x be the distance between the child and the light, and t be the time in seconds. You are given that the rate at which the child is moving (dx/dt) is 1 m/s. You are asked to determine the rate of change of the shadow (dS/dt) when the child is 10 m from the building ($x = 40$ m).

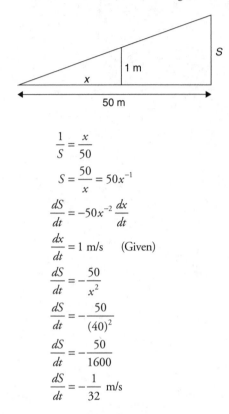

$$\frac{1}{S} = \frac{x}{50}$$

$$S = \frac{50}{x} = 50x^{-1}$$

$$\frac{dS}{dt} = -50x^{-2}\frac{dx}{dt}$$

$$\frac{dx}{dt} = 1 \text{ m/s} \quad \text{(Given)}$$

$$\frac{dS}{dt} = -\frac{50}{x^2}$$

$$\frac{dS}{dt} = -\frac{50}{(40)^2}$$

$$\frac{dS}{dt} = -\frac{50}{1600}$$

$$\frac{dS}{dt} = -\frac{1}{32} \text{ m/s}$$

155. (D) Let x be the height of the rocket in meters, θ be the camera angle in degrees, y be the distance from the camera to the rocket, and t be the time in seconds. You are given the velocity of the rocket ($dx/dt = 20$ m/s).

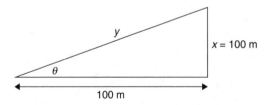

$$\text{Tan } \theta = \frac{x}{100}$$

$$\sec^2 \theta \frac{d\theta}{dt} = \frac{1}{100} \frac{dx}{dt}$$

$$\frac{dx}{dt} = 20 \text{ m/s}$$

$$\frac{d\theta}{dt} = \left(\frac{1}{100}\right)\left(\frac{1}{\sec^2 \theta}\right)(20)$$

$$\frac{d\theta}{dt} = \frac{1}{5\sec^2 \theta}$$

$$y^2 = x^2 + (100)^2 \quad \text{Pythagorean theorem}$$

$$y^2 = (100)^2 + (100)^2$$

$$y^2 = 2(100)^2$$

$$y = 100\sqrt{2}$$

$$\sec \theta = \frac{y}{100} \quad \text{at } x = 100 \text{ m}$$

$$\sec \theta = \frac{100\sqrt{2}}{100} = \sqrt{2}$$

$$\frac{d\theta}{dt} = \frac{1}{5\sec^2 \theta}$$

$$\frac{d\theta}{dt} = \left(\frac{1}{5}\right)\frac{1}{(\sqrt{2})^2}$$

$$\frac{d\theta}{dt} = \left(\frac{1}{5}\right)\left(\frac{1}{2}\right)$$

$$\frac{d\theta}{dt} = \frac{1}{10} \text{ rad/s} = 5.7 \text{ deg/s}$$

156. (D) Let l be the length, w be the width, and A be the area. You are given that the rate of increase of the width (dw/dt) is twice that of the length (dl/dt). You are asked to find the ratio of the rate of change of the area (dA/dt) to that of the width when the length is twice the width ($l = 2w$).

$$A = lw$$
$$\frac{dA}{dt} = l\frac{dw}{dt} + w\frac{dl}{dt}$$
$$l = 2w$$
$$\frac{dw}{dt} = 2\frac{dl}{dt}$$
$$\frac{dl}{dt} = \frac{1}{2}\frac{dw}{dt}$$
$$\frac{dA}{dt} = (2w)\frac{dw}{dt} + w\left(\frac{1}{2}\right)\left(\frac{dw}{dt}\right)$$
$$\frac{dA}{dt} = (2.5w)\frac{dw}{dt}$$

157. (B) You are given the cost function $C(x) = 200 + 4x + 0.5x^2$ and $x = 50$. The marginal cost is the first derivative of the cost function.

$$C(x) = 200 + 4x + 0.5x^2$$
$$C'(x) = 4 + x$$
$$x = 50$$
$$C'(x) = 4 + 50$$
$$C'(x) = \$54$$

158. (B) The position function of the particle with time in seconds is $f(x) = x^3 - 6x^2 + 9x$. The velocity is the first derivative of the function and the times are when the first derivative is equal to zero.

$$f(x) = x^3 - 6x^2 + 9x$$
$$f'(x) = 3x^2 - 12x + 9$$
$$0 = 3x^2 - 12x + 9$$
$$0 = x^2 - 4x + 3$$
$$0 = (x - 1)(x - 3)$$
$$x - 1 = 0 \qquad x - 3 = 0$$
$$x = 1\text{ s} \qquad x = 3\text{ s}$$

159. (E) You are given that the perimeter (P) of the fence is 500 ft and the shape must be a rectangle. Let l be the length, w be the width, and A be the area. You can calculate the dimensions for maximum area.

$$p = 2l + 2w$$
$$500 = 2l + 2w$$
$$250 = l + w$$
$$l = 250 - w$$

$$A = lw$$
$$A(w) = (250 - w)w$$
$$A(w) = 250w - w^2$$
$$A'(w) = 250 - 2w$$
$$0 = 250 - 2w$$
$$0 = 125 - w$$
$$w = 125 \text{ ft}$$
$$l = (250 - w) = (250 - 125) = 125 \text{ ft}$$

160. (B) You know the man and the woman's speed ($dx/dt = dy/dt = 1$ m/s) and their distances from the intersection ($x = y = 100$ m). You can find the speed (dz/dt) at which they are separating.

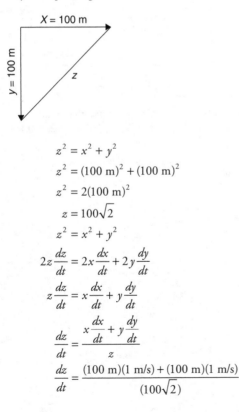

$$z^2 = x^2 + y^2$$
$$z^2 = (100 \text{ m})^2 + (100 \text{ m})^2$$
$$z^2 = 2(100 \text{ m})^2$$
$$z = 100\sqrt{2}$$
$$z^2 = x^2 + y^2$$
$$2z\frac{dz}{dt} = 2x\frac{dx}{dt} + 2y\frac{dy}{dt}$$
$$z\frac{dz}{dt} = x\frac{dx}{dt} + y\frac{dy}{dt}$$
$$\frac{dz}{dt} = \frac{x\dfrac{dx}{dt} + y\dfrac{dy}{dt}}{z}$$
$$\frac{dz}{dt} = \frac{(100 \text{ m})(1 \text{ m/s}) + (100 \text{ m})(1 \text{ m/s})}{(100\sqrt{2})}$$

$$\frac{dz}{dt} = \frac{(2)(100 \text{ m})(1 \text{ m/s})}{(100\sqrt{2})}$$

$$\frac{dz}{dt} = \frac{2}{\sqrt{2}} \text{ m/s}$$

$$\frac{dz}{dt} = 1.4 \text{ m/s}$$

161. (B) You must take the first derivative of $f(x) = 2x^2 - 4x + 2$ and evaluate it at $x = 20$.

$$f(x) = 2x^2 - 4x + 2$$
$$f'(x) = 4x - 4$$
$$f'(20) = 4(20) - 4$$
$$f'(20) = 76$$

162. (A) The instantaneous acceleration is the second derivative of $f(x) = 2x^3 - 4x^2 + 2x - 5$ evaluated at $x = 10$.

$$f(x) = 2x^3 - 4x^2 + 2x - 5$$
$$f'(x) = 6x^2 - 8x + 2$$
$$f''(x) = 12x - 8$$
$$f''(10) = 12(10) - 8$$
$$f''(10) = 112 \text{ m/s}^2$$

163. (D) You must first set up the equation for profit, differentiate it, and evaluate it for $x = 1000$ units. The price of each unit is $500 and the cost function is $C(x) = 200 + 16x + 0.1x^2$.

$$P = R - C$$
$$R = xp = 500x$$
$$C(x) = 200 + 16x + 0.1x^2$$
$$P = 500x - (200 + 16x + 0.1x^2)$$
$$P = 500x - 200 - 16x - 0.1x^2$$
$$P = -200 + 484x - 0.1x^2$$
$$\frac{dP}{dx} = 484 - 0.2x$$
$$x = 1000$$
$$\frac{dP}{dx} = 484 - 0.2(1000)$$
$$\frac{dP}{dx} = 484 - 200$$
$$\frac{dP}{dx} = \$284$$

164. (E) A right cylindrical cone has a radius of 4 cm and a height of 2.0 cm. The height increases ($dh/dt = 0.5$ cm/min), but the radius remains constant. You can calculate the rate of change of volume.

$$V = \frac{1}{3}\pi r^2 h$$

$$\frac{dV}{dt} = \frac{1}{3}\pi r^2 \frac{dh}{dt}$$

$$\frac{dh}{dt} = 0.5 \text{ cm/min}, \ r = 4 \text{ cm constant}$$

$$\frac{dV}{dt} = \frac{1}{3}\pi (4 \text{ cm})^2 (0.5 \text{ cm/min})$$

$$\frac{dV}{dt} = 8.4 \text{ cm}^3/\text{min}$$

165. (D) To find the coordinates of the vertex of the parabola $y = 3x^2 + 2$, take the first derivative of the function. To find the x-coordinate, set the first derivative $= 0$ and solve for x. To find the y-coordinate, evaluate the function at $x = 0$.

$$y = 3x^2 + 2$$

$$\frac{dy}{dx} = 6x$$

$$0 = 6x$$

$$x = 0$$

$$y = 3x^2 + 2$$

$$y = 3(0)^2 + 2$$

$$y = 2$$

The coordinates of the vertex are $(0, 2)$.

166. (D) Let x be the amount of water pumped in, y be the amount of water removed, h be the level of the water in the tank, and V be the volume of water in the tank. The radius of the tank is 10 m. The rate of water pumped in (dx/dt) is 100 m³/min and the rate of water removed (dy/dt) is 70 m³/min. You can calculate the approximate rate of change of the water level in the tank.

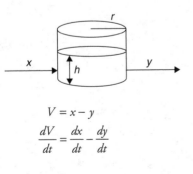

$$V = x - y$$

$$\frac{dV}{dt} = \frac{dx}{dt} - \frac{dy}{dt}$$

$$V = \pi r^2 h$$

$$\frac{dV}{dt} = \pi r^2 \frac{dh}{dt} \quad (r \text{ is constant})$$

$$\pi r^2 \frac{dh}{dt} = \frac{dx}{dt} - \frac{dy}{dt}$$

$$\frac{dh}{dt} = \frac{1}{\pi r^2} \left(\frac{dx}{dt} - \frac{dy}{dt} \right)$$

$$\frac{dh}{dt} = \frac{1}{\pi (10 \text{ m})^2} ((100 \text{ m}^3/\text{min}) - (70 \text{ m}^3/\text{min}))$$

$$\frac{dh}{dt} = \left(\frac{1}{314 \text{ m}^2} \right) (300 \text{ m}^3/\text{min})$$

$$\frac{dh}{dt} = 0.1 \text{ m/min}$$

167. (C) The ladder ($z = 8$ ft), the height of the top of the ladder against the wall (y) and the distance of the bottom of the ladder from the wall along the ground (x) form a right triangle. You are given the rate of fall of the top of the ladder ($dy/dt = -1$ ft/s) at $y = 4$ ft. You can calculate the rate that the bottom of the ladder moves with the Pythagorean theorem and its derivative.

$$x^2 + y^2 = z^2$$

$$x^2 + y^2 = (8)^2$$

$$x^2 + y^2 = 64$$

$$2x\frac{dx}{dt} + 2y\frac{dy}{dt} = 0$$

$$x\frac{dx}{dt} + y\frac{dy}{dt} = 0$$

$$x\frac{dx}{dt} = -y\frac{dy}{dt}$$

$$\frac{dx}{dt} = -\frac{y}{x}\frac{dy}{dt}$$

$$x^2 + (4)^2 = 64$$

$$x^2 + 16 = 64$$

$$x^2 = 48$$

$$x = \sqrt{48} = 4\sqrt{3}$$

$$\frac{dx}{dt} = -\frac{(4 \text{ ft})}{(4\sqrt{3} \text{ ft})}(-1 \text{ ft/s})$$

$$\frac{dx}{dt} = \frac{1}{\sqrt{3}} \text{ ft/s} = \frac{\sqrt{3}}{3} \text{ ft/s}$$

168. (B) You are given the function $Q = mc\Delta T$, mass of iron block (1 kg), the specific heat capacity of iron ($c = 460$ J/kg · K), and the rate of thermal energy input ($dQ/dt = 100$ J/s). You are asked to find the rate of temperature rise ($d(\Delta T)/dt$).

$$Q = mc\Delta T$$

$$\frac{dQ}{dt} = mc\frac{d(\Delta T)}{dt}$$

$$\frac{d(\Delta T)}{dt} = \frac{1}{mc}\frac{dQ}{dt}$$

$$\frac{d(\Delta T)}{dt} = \frac{1}{((1\text{kg})(460 \text{ J/kg}\cdot\text{K})}(100 \text{ J/s})$$

$$\frac{d(\Delta T)}{dt} = 0.22 \text{ K/s}$$

169. (E) The rate of change of the balloon's volume (dV/dt) is 50 cm³/s and the radius of the balloon (r) is 10 cm. Let D be the balloon's diameter. You can find the rate of change of the balloon's diameter (dD/dt).

$$V = \frac{4}{3}\pi r^3$$

$$\frac{dV}{dt} = 4\pi r^2 \frac{dr}{dt}$$

$$D = 2r$$

$$\frac{dD}{dt} = 2\frac{dr}{dt}$$

$$\frac{dr}{dt} = \frac{1}{2}\frac{dD}{dt}$$

$$\frac{dV}{dt} = 4\pi r^2\left(\frac{1}{2}\frac{dD}{dt}\right)$$

$$\frac{dV}{dt} = 2\pi r^2 \frac{dD}{dt}$$

$$\frac{dD}{dt} = \frac{\left(\dfrac{dV}{dt}\right)}{2\pi r^2}$$

$$\frac{dD}{dt} = \frac{(50 \text{ cm}^3/\text{s})}{2\pi(10 \text{ cm})^2}$$

$$\frac{dD}{dt} = \frac{(50 \text{ cm}^3/\text{s})}{2\pi(100 \text{ cm}^2)}$$

$$\frac{dD}{dt} = \frac{1}{4\pi} \text{ cm/s}$$

170. (E) The length (L) of the farmer's fence is 160 m. Let x be the length of the rectangle and y be the width. Now use the area formula to solve the problem.

$$L = 2x + y$$
$$160 = 2x + y$$
$$y = 160 - 2x$$
$$A = xy$$
$$A = x(160 - 2x)$$
$$A = 160x - 2x^2$$
$$\frac{dA}{dt} = 160 - 4x$$
$$0 = 160 - 4x$$
$$4x = 160$$
$$x = 40 \text{ m}$$
$$y = 160 - 2(40 \text{ m}) = 80 \text{ m}$$
$$A = (40 \text{ m})(80 \text{ m})$$
$$A = 3200 \text{ m}^2$$

171. (C) Given the cost function $C(x) = 144 + 0.1x + 0.04x^2$, you can find the minimum average cost per unit.

$$C(x) = 144 + 0.1x + 0.04x^2$$
$$\overline{C} = \frac{C}{x} = \frac{144}{x} + 0.1 + 0.04x$$
$$\overline{C}' = -144x^{-2} + 0.04$$
$$0 = -\frac{144}{x^2} + 0.04$$
$$\frac{144}{x^2} = 0.04$$
$$144 = 0.04x^2$$
$$x^2 = \frac{144}{0.04} = 3600$$
$$x = \$60$$

172. (B) Let $p(x, y)$ be the vertex of the rectangle on the line. Use the area formula and differentiate it. Set the first derivative to zero and find the critical number. Then substitute the x value into the line's equation to find the dimensions.

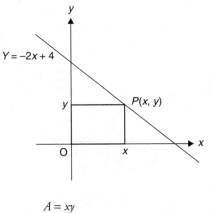

$Y = -2x + 4$

$P(x, y)$

$A = xy$

$A = x(-2x + 4)$ The domain of A is $[0, 2]$

$A = -2x^2 + 4x$

$\dfrac{dA}{dx} = -4x + 4$

$0 = -4x + 4$

$4x = 4$

$x = 1$ length

$y = -2(1) + 4 = 2$ width

173. (E) Find two negative numbers whose sum is -50 so that they have the maximum product. Let y be the product of the two numbers.

$y = -x(-50 + x)$

$y = 50x + x^2$

$\dfrac{dy}{dx} = 50 + 2x$

$0 = 50 + 2x$

$2x = -50$

$x = -25$

174. (D) Let V be the volume of the sphere, S be the surface area of the sphere, and r be the radius of the sphere. You are given that $V = (1/(48\pi^2))$ cubic units. You must calculate the radius of the sphere, differentiate the surface area equation, and substitute the radius of the sphere into that equation.

$V = \dfrac{4}{3}\pi r^3$

$\dfrac{1}{48\pi^2} = \dfrac{4}{3}\pi r^3$

$$1 = 64\pi^3 r^3$$

$$r^3 = \frac{1}{64\pi^3}$$

$$r = \frac{1}{4\pi}$$

$$S = 4\pi r^2$$

$$\frac{dS}{dt} = 8\pi r \frac{dr}{dt}$$

$$\frac{dS}{dt} = 8\pi \left(\frac{1}{4\pi}\right) \frac{dr}{dt}$$

$$\frac{dS}{dt} = 2\frac{dr}{dt}$$

175. (C) Let x be the length of the square. So, the length of the box will be $8 - 2x$ and the width will be $4 - 2x$. The range of x must be $0 \le x \le 2$ cm. You differentiate the volume formula to find the maximum size of x.

$$V = lwh$$

$$V = x(4 - 2x)(8 - 2x)$$

$$V = 4x^3 - 24x^2 + 32x$$

$$\frac{dV}{dx} = 12x^2 - 48x + 32$$

$$0 = 12x^2 - 48x + 32$$

$$0 = 3x^2 - 12x + 8$$

By the quadratic formula:

$$x = 3.15 \text{ and } x = 0.845$$

However, $x = 3.15$ is out of the range of x

$$x = 0.845 \text{ cm}$$

176. (D) You are given the cost function $C(x) = 500 + 6x + 0.2x^2$ and the price function $p(x) = 20$. Calculate the revenue and the profit. Differentiate the profit function and find the critical number, which will be the number of units for maximum profit.

$$R(x) = xp(x) = 20x \quad \text{(Calculate revenue)}$$

$$C(x) = 500 + 6x + 0.2x^2 \quad \text{(Given cost function)}$$

$$P(x) = R(x) - C(x) \quad \text{(Profit equation)}$$

$$P(x) = 20x - (500 + 6x + 0.2x^2)$$

$$P(x) = 20x - 500 - 6x - 0.2x^2$$

$$P(x) = -500 + 14x - 0.2x^2$$

$$P'(x) = 14 - 0.4x$$
$$0 = 14 - 0.4x$$
$$0.4x = 14$$
$$x = 35 \text{ units}$$

177. (D) Find two numbers whose sum is 30 so that they have the maximum product. Let y be the product of the two numbers.

$$y = x(30 - x)$$
$$y = 30x - x^2$$
$$\frac{dy}{dx} = 30 - 2x$$
$$0 = 30 - 2x$$
$$2x = 30$$
$$x = 15$$

178. (D) You know the speeds of car A and car B ($dx/dt = 100$ km/h, $dy/dt = 50$ km/h). After 1 h, you can calculate their distances from the intersection ($x = 100$ km, $y = 50$ km). You can find the speed (dz/dt) at which they are separating.

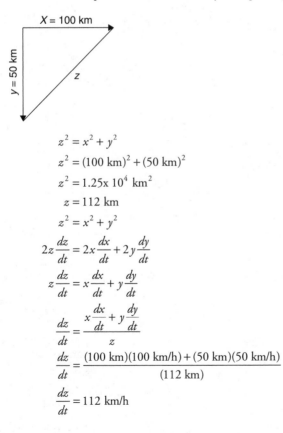

X = 100 km

y = 50 km

z

$$z^2 = x^2 + y^2$$
$$z^2 = (100 \text{ km})^2 + (50 \text{ km})^2$$
$$z^2 = 1.25 \text{x } 10^4 \text{ km}^2$$
$$z = 112 \text{ km}$$
$$z^2 = x^2 + y^2$$
$$2z\frac{dz}{dt} = 2x\frac{dx}{dt} + 2y\frac{dy}{dt}$$
$$z\frac{dz}{dt} = x\frac{dx}{dt} + y\frac{dy}{dt}$$
$$\frac{dz}{dt} = \frac{x\frac{dx}{dt} + y\frac{dy}{dt}}{z}$$
$$\frac{dz}{dt} = \frac{(100 \text{ km})(100 \text{ km/h}) + (50 \text{ km})(50 \text{ km/h})}{(112 \text{ km})}$$
$$\frac{dz}{dt} = 112 \text{ km/h}$$

179. (E) You are given the function $Q = mc\Delta T$, mass of copper block (1 kg), the specific heat capacity of copper ($c = 390$ J/kg · K), and the rate of temperature rise ($d(\Delta T)/dt = 0.026$ K/s). You are asked to find the rate of thermal energy input (dQ/dt).

$$Q = mc\Delta T$$

$$\frac{dQ}{dt} = mc\frac{d(\Delta T)}{dt}$$

$$\frac{dQ}{dt} = (1 \text{ kg})(390 \text{ J/kg} \cdot \text{K})(0.026 \text{ K/s})$$

$$\frac{dQ}{dt} = 10 \text{ J/s}$$

180. (D) The length of the rope is 90 ft. Let h be the height of the piano above the ground, y be the length of rope from the pulley to the piano, x be the horizontal distance between the man and the piano, and z be the length of rope from the pulley to the man's hand. The pulley is 40 ft above the man's arms. At $t = 0$, the man is 30 ft horizontally from the piano and walks away at 12 ft/s. Find how fast the piano is being pulled up.

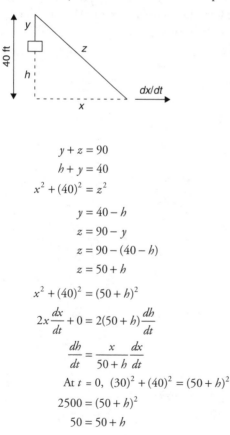

$$y + z = 90$$

$$h + y = 40$$

$$x^2 + (40)^2 = z^2$$

$$y = 40 - h$$

$$z = 90 - y$$

$$z = 90 - (40 - h)$$

$$z = 50 + h$$

$$x^2 + (40)^2 = (50 + h)^2$$

$$2x\frac{dx}{dt} + 0 = 2(50 + h)\frac{dh}{dt}$$

$$\frac{dh}{dt} = \frac{x}{50 + h}\frac{dx}{dt}$$

At $t = 0$, $(30)^2 + (40)^2 = (50 + h)^2$

$$2500 = (50 + h)^2$$

$$50 = 50 + h$$

$$\frac{dh}{dt} = \frac{x}{50+h} \frac{dx}{dt}$$

$$\frac{dh}{dt} = \frac{30}{50}(12)$$

$$\frac{dh}{dt} = \frac{36}{5} = 5.2 \text{ ft/s}$$

181. (D) The position function of the particle with time in seconds is $f(x) = x^3 - 9x^2 + 24x$. Acceleration is the second derivative of the function and the time is when the second derivative is equal to zero.

$$f(x) = x^3 - 9x^2 + 24x$$
$$f'(x) = 3x^2 - 18x + 24$$
$$f''(x) = 6x - 18$$
$$0 = 6x - 18$$
$$6x = 18$$
$$x = 3 \text{ s}$$

182. (D) The width of the metal used to make the gutter is 16 in. The maximum capacity will occur when the rectangle seen end on has the greatest area. The dimensions of the rectangle are shown. You must differentiate the area formula and find a maximum.

$$f(x) = x(16 - 2x) \qquad 0 \le x \le 8$$
$$f(x) = 16x - 2x^2$$
$$f'(x) = 16 - 4x$$
$$0 = 16 - 4x$$
$$4x = 16$$
$$x = 4 \text{ in}$$

183. (D) The ellipsoid has the following radii: $r_a = 8$ cm, $r_b = 4$ cm, and $r_c = 2$ cm. r_a remains constant, but r_b increases by 0.5 cm/min and r_c increases by 2 cm/min. Given the formula for the volume of an ellipsoid, you must differentiate it and evaluate it with the given values.

$$V = \frac{4}{3}\pi r_a r_b r_c$$

$$\frac{dV}{dt} = \left(\frac{4}{3}\pi r_a\right)\left(r_c \frac{dr_b}{dt} + r_b \frac{dr_c}{dt}\right) \quad r_a \text{ is constant}$$

$$\frac{dV}{dt} = \left(\frac{4}{3}\pi(8\ \text{cm})\right)((2\ \text{cm})(0.5\ \text{cm/min}) + (4\ \text{cm})(2\ \text{cm/min}))$$

$$\frac{dV}{dt} = \left(\frac{32\pi}{3}\ \text{cm}\right)(9\ \text{cm}^2/\text{min})$$

$$\frac{dV}{dt} = 96\pi\ \text{cm}^3/\text{min}$$

184. (E) The angle of the plane's take-off is 30°. Let h be the plane's altitude and t be time in hours. The plane's velocity is 500 mil/h. You can find the rate of change of the plane's altitude.

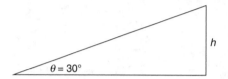

$$\sin(30°) = \frac{h}{500t}$$

$$h = \sin(30°)500t$$

$$\frac{dh}{dt} = \sin(30°)500$$

$$\frac{dh}{dt} = 250\ \text{mil/h}$$

185. (D) You must first set up the equation for profit and then differentiate it, and evaluate it for $x = 500$ units. The price of each unit is $500 and the cost function is $C(x) = 300 + 4x + 0.2x^2$.

$$P(x) = R(x) - C(x)$$

$$R(x) = xp = 500x$$

$$C(x) = 300 + 4x + 0.2x^2$$

$$P(x) = 500x - (300 + 4x + 0.2x^2)$$

$$P(x) = 500x - 300 - 4x - 0.2x^2$$

$$P(x) = -300 + 496x - 0.2x^2$$

$$P'(x) = 496 - 0.4x$$

186. (D) The length (L) of the fence is 600 m. Let x be the length of the rectangle and y be the width. Now use the area formula to solve the problem.

$$L = 2x + y$$

$$600 = 2x + y$$

$$y = 600 - 2x$$

$$A = xy$$
$$A = x(600 - 2x)$$
$$A = 600x - 2x^2$$
$$\frac{dA}{dt} = 600 - 4x$$
$$0 = 600 - 4x$$
$$4x = 600$$
$$x = 150 \text{ m}$$
$$y = 600 - 2(150 \text{ m}) = 300 \text{ m}$$
$$A = (150 \text{ m})(300 \text{ m})$$
$$A = 45000 \text{ m}^2$$

187. (D) Let x be the amount of oil pumped in, y be the amount of oil removed, h be the level of the oil in the tank, and V be the volume of oil in the tank. The radius of the tank is 100 m. The rate of oil pumped in (dx/dt) is 5000 m³/min and the rate of oil removed (dy/dt) is 100 m³/min. You can calculate the approximate rate of change of the oil level in the tank.

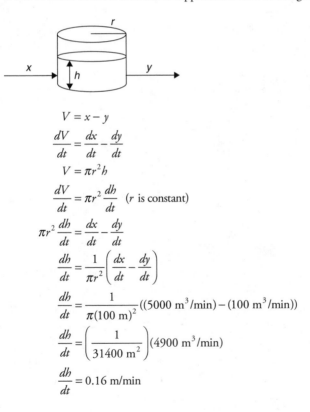

$$V = x - y$$
$$\frac{dV}{dt} = \frac{dx}{dt} - \frac{dy}{dt}$$
$$V = \pi r^2 h$$
$$\frac{dV}{dt} = \pi r^2 \frac{dh}{dt} \quad (r \text{ is constant})$$
$$\pi r^2 \frac{dh}{dt} = \frac{dx}{dt} - \frac{dy}{dt}$$
$$\frac{dh}{dt} = \frac{1}{\pi r^2}\left(\frac{dx}{dt} - \frac{dy}{dt}\right)$$
$$\frac{dh}{dt} = \frac{1}{\pi(100 \text{ m})^2}((5000 \text{ m}^3/\text{min}) - (100 \text{ m}^3/\text{min}))$$
$$\frac{dh}{dt} = \left(\frac{1}{31400 \text{ m}^2}\right)(4900 \text{ m}^3/\text{min})$$
$$\frac{dh}{dt} = 0.16 \text{ m/min}$$

188. (C) The instantaneous velocity is the first derivative of $f(x) = 16x^3 - 14x^2 + 6x - 7$ evaluated at $x = 0.5$.

$$f(x) = 16x^3 - 14x^2 + 6x - 7$$
$$f'(x) = 48x^2 - 28x + 6$$
$$f'(0.5) = 48(0.5)^2 - 28(0.5) + 6$$
$$f'(0.5) = 12 - 14 + 6$$
$$f'(0.5) = 4 \text{ m/s}$$

189. (E) The cone has an altitude of 15 cm and a base radius of 3 cm. The cylinder has a radius of r and a height of h. Write h in terms of r and substitute it into the volume of a cylinder formula. Differentiate the volume of a cylinder equation, set the derivative to zero, and find the critical values. Once found, use the critical values to find the volume of the cylinder.

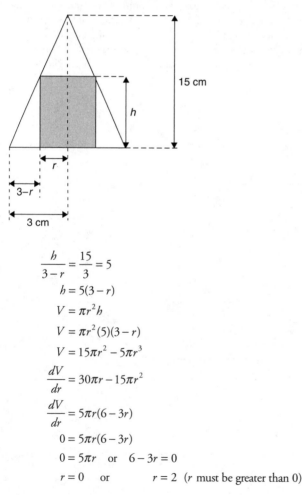

$$\frac{h}{3-r} = \frac{15}{3} = 5$$
$$h = 5(3-r)$$
$$V = \pi r^2 h$$
$$V = \pi r^2 (5)(3-r)$$
$$V = 15\pi r^2 - 5\pi r^3$$
$$\frac{dV}{dr} = 30\pi r - 15\pi r^2$$
$$\frac{dV}{dr} = 5\pi r(6 - 3r)$$
$$0 = 5\pi r(6 - 3r)$$
$$0 = 5\pi r \quad \text{or} \quad 6 - 3r = 0$$
$$r = 0 \quad \text{or} \quad r = 2 \ (r \text{ must be greater than } 0)$$

$$h = 5(3-2) = 5$$
$$V = \pi(2)^2(5)$$
$$V = 20\pi \text{ cm}^3$$

190. (A) The pressure across a bubble (Δp) is given by this equation: $\Delta p = 2\gamma/r$, where γ is the surface tension and r is the radius of the bubble. As the pressure across the bubble increases, the bubble shrinks. To calculate how the rate of increase in pressure change compares to the rate of change of the radius, you must first differentiate the surface area equation. Use the given ratio of the rate of change of the surface area to the rate of decrease of the radius of the bubble to calculate the radius of the bubble. Differentiate the pressure equation (assume that γ is constant) and then substitute the radius into the differentiated pressure equation.

$$S = 4\pi r^2$$
$$\frac{dS}{dt} = 8\pi r \frac{dr}{dt}$$
$$\frac{dS}{dt} = 4\frac{dr}{dt} \qquad \text{(Given)}$$
$$8\pi r \frac{dr}{dt} = 4\frac{dr}{dt}$$
$$2\pi r = 1$$
$$r = \frac{1}{2\pi}$$
$$\Delta p = \frac{2\gamma}{r}$$
$$\Delta p = 2\gamma r^{-1}$$
$$\frac{d(\Delta p)}{dt} = -2\gamma r^{-2}\frac{dr}{dt}$$
$$\frac{d(\Delta p)}{dt} = -\frac{2\gamma}{r^2}\frac{dr}{dt}$$
$$\frac{d(\Delta p)}{dt} = -\frac{2\gamma}{\left(\dfrac{1}{2\pi}\right)^2}\frac{dr}{dt}$$
$$\frac{d(\Delta p)}{dt} = (-8\pi^2\gamma)\frac{dr}{dt} \text{ pressure units/time}$$

191. (C) The base of the can is 10 cm, so the radius(r) is 5.0 cm. The height of the can is 10 cm. The rate of the drainage (dV/dt) is a constant 4 cm³/s. So, what is the change in height of the water level (dh/dt)?

$$V = \pi r^2 h$$
$$\frac{dV}{dt} = \pi r^2 \frac{dh}{dt} \qquad (r \text{ is constant})$$

$$\frac{dh}{dt} = \left(\frac{1}{\pi r^2}\right)\frac{dV}{dt}$$

$$\frac{dh}{dt} = \left(\frac{1}{\pi(5 \text{ cm})^2}\right)(-4 \text{ cm}^3)$$

$$\frac{dh}{dt} = -\frac{4}{25\pi} \text{ cm/s}$$

192. (A) You are given the cost function $C(x) = x^2 + 20x + 4$. Divide this function by x to get the average, differentiate the average function, set the derivative to zero, and find the minimum.

$$C(x) = x^2 + 20x + 4$$

$$\overline{C}(x) = \frac{x^2 + 20x + 4}{x}$$

$$\overline{C}(x) = x + 20 + \frac{4}{x}$$

$$\overline{C}'(x) = 1 - \frac{4}{x^2}$$

$$0 = 1 - \frac{4}{x^2}$$

$$0 = x^2 - 4$$

$$x^2 = 4$$

$$x = \pm 2$$

$$x = 2 \quad (x \text{ must be greater than } 0)$$

193. (E) Acceleration is the first derivative of velocity. So, take the derivative of the given velocity function, set the expression equal to zero, and solve for the time or times.

$$v = \frac{1}{3}t^3 - 2t^2 + 3t + 2$$

$$\frac{dv}{dt} = t^2 - 4t + 3$$

$$0 = t^2 - 4t + 3$$

$$0 = (t - 1)(t - 3)$$

$$t - 1 = 0 \qquad t - 3 = 0$$

$$t = 1 \text{ s} \qquad t = 3 \text{ s}$$

194. (D) The height of a football when punted into the air is given by the function: $y(t) = v_0 t - \frac{1}{2}gt^2$. The initial velocity of the football (v_0) is 20 m/s, g is acceleration due to gravity (10 m/s²), and t is time in seconds. You can calculate how long it will take to reach its maximum height.

$$y(t) = v_0 t - \frac{1}{2}gt^2$$
$$y'(t) = v_0 - gt$$
$$0 = v_0 - gt$$
$$gt = v_0$$
$$t = \frac{v_0}{g}$$
$$t = \frac{(20 \text{ m/s})}{(10 \text{ m/s}^2)}$$
$$t = 2 \text{ s}$$

195. (E) You are given a position function: $f(x) = 2x^2 - 20x + 5$ in meters. The acceleration function is the second derivative of $f(x)$ and its unit is m/s².

$$f(x) = 2x^2 - 20x + 5$$
$$f'(x) = 4x - 20$$
$$f''(x) = 4$$

So, acceleration = 4 m/s²

196. A student builds an experimental model rocket with a variable thrust engine. The test flight lasted for 5 s and the rocket's altitude (in meters) could be described by the following function:

$$y(t) = \frac{1}{3}t^3 - 2t^2 + 3t + 2$$

(A) What is the velocity function of the rocket (m/s)? What is the acceleration function of your rocket (m/s²)?

Velocity is the first derivative of the altitude function: $y'(t) = t^2 - 4t + 3$.
Acceleration is the second derivative of the altitude function: $y''(t) = 2t - 4$.

(B) Make graphs of each function vs. time (s) for 0 to 5 s. Derive the maxima and minima for each graph (show your work).

Altitude function:

Time (s)	Altitude (m) = y(t)
0	2.0
1	3.3
2	2.6
3	1.9
4	3.1

Critical points:

$$y'(t) = x^2 - 4x + 3$$
$$0 = x^2 - 4x + 3 = (x-1)(x-3)$$

$x - 1 = 0 \rightarrow x = 1$ $y(1) = \frac{1}{3}(1)^3 - 2(1)^2 + 3(1) + 2 = 3.3$ maximum (1, 3.3)

$x - 3 = 0 \rightarrow x = 3$ $y(3) = \frac{1}{3}(3)^3 - 2(3)^2 + 3(3) + 2 = 1.9$ minimum (3, 1.9)

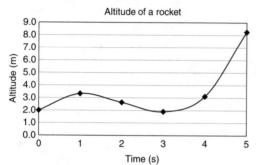

Velocity function:

Time (s)	Velocity (m/s) = $y'(t)$
0	3
1	0
2	−1
3	0
4	3

Critical points:

$$y''(t) = 2x - 4$$
$$0 = 2x - 4$$

$2x - 4 = 0 \rightarrow x = 2$ $y'(2) = (2)^2 - 4(2)^2 + 3 = -1$ minimum (2, −1)

$y'(0) = (0)^2 - 4(0)^2 + 3 = 3$

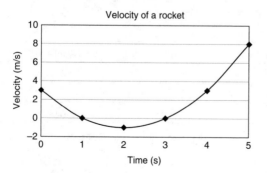

Acceleration function:

Time (s)	Acceleration (m/s²) = $y''(t)$
0	3
1	0
2	−1
3	0
4	3

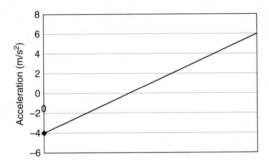

197. A conical funnel has a base diameter of 6 cm and a height of 5 cm. The funnel sits over a cylindrical can with an open top. The can has a diameter of 4 cm and a height of 5 cm. The funnel is initially full, but water is draining from the funnel bottom into the can at a constant rate of 2 cm³/s. Answer the following questions (show your work):

(A) How fast is the water level in the funnel falling when the water is 2.5 cm high?

Let's look at the funnel first.

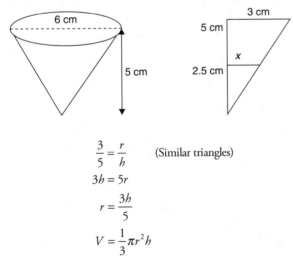

$$\frac{3}{5} = \frac{r}{h} \qquad \text{(Similar triangles)}$$

$$3h = 5r$$

$$r = \frac{3h}{5}$$

$$V = \frac{1}{3}\pi r^2 h$$

$$V = \frac{1}{3}\pi\left(\frac{3h}{5}\right)^2 h$$

$$V = \frac{1}{3}\pi\left(\frac{9h^2}{25}\right)h$$

$$V = \frac{3}{25}\pi h^3$$

$$\frac{dV}{dt} = 3\left(\frac{3}{25}\right)\pi h^2\frac{dh}{dt}$$

$$\frac{dV}{dt} = \left(\frac{9}{75}\right)\pi h^2\frac{dh}{dt}$$

$$\frac{dV}{dt} = -2 \text{ cm}^3/s$$

$$-2 = \left(\frac{9}{75}\right)\pi h^2\frac{dh}{dt}$$

$$\frac{dh}{dt} = -\frac{75}{9\pi h^2}$$

$$\frac{dh}{dt} = -\frac{75}{9\pi(2.5)^2}$$

$$\frac{dh}{dt} = -\frac{75}{56\pi} \text{ cm/s} = -177 \text{ cm/s}$$

(B) How fast is the water level in the can rising?

Now let's look at the can. r is the radius, which is constant at 2 cm. Let x be the amount of water flowing into the can. The rate at which water flows into the can is equal to the rate at which water is draining from the funnel (-2 cm^3/s).

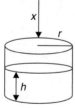

$$V = x$$

$$\frac{dV}{dt} = \frac{dx}{dt}$$

$$V = \pi r^2 h$$

$$\frac{dV}{dt} = \pi r^2\frac{dh}{dt} \quad (r \text{ is constant})$$

$$\pi r^2 \frac{dh}{dt} = \frac{dx}{dt}$$

$$\frac{dh}{dt} = \frac{1}{\pi r^2}\left(\frac{dx}{dt}\right)$$

$$\frac{dh}{dt} = \frac{1}{\pi(2 \text{ cm})^2}(2 \text{ cm}^3/\text{min})$$

$$\frac{dh}{dt} = \left(\frac{1 \text{ cm/s}}{2\pi}\right) = 0.16 \text{ cm/s}$$

(C) Will the can overflow? If not, how high will the final level of water be in the can? If the volume of the funnel is greater than the volume of the cylinder, then the can will overflow.

Cone

$$V = \frac{1}{3}\pi r^2 h$$

$$V = \frac{1}{3}\pi(3 \text{ cm})^2(5 \text{ cm})$$

$$V = 15\pi \text{ cm}^3 = 47 \text{ cm}^3$$

∴ The cylinder will not overflow.

Cylinder

$$V = \pi r^2 h$$

$$V = \pi(2 \text{ cm})^2(5 \text{ cm})$$

$$V = 20\pi \text{ cm}^3 = 63 \text{ cm}^3$$

Water level in the cylinder

$$V = \pi r^2 h$$

$$h = \frac{V}{\pi r^2} = \frac{(15\pi \text{ cm}^3)}{\pi(5 \text{ cm})^2} = 3 \text{ cm}$$

198. The perimeter of an isosceles triangle is 16 cm. Do the following:

(A) Make a drawing of the problem.

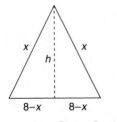

Given: $P = 16$ cm

(B) What are the dimensions of the sides and height for the maximum area?

$$x^2 = h^2 + (8 - x)^2$$

$$x^2 = h^2 + (64 - 16x + x^2)$$

$$x^2 = h^2 + 64 - 16x + x^2$$

$$0 = h^2 + 64 - 16x$$

$$h^2 = 16x - 64$$

$$h^2 = 16(x - 4)$$

$$h = 4\sqrt{(x - 4)}$$

$$A = \frac{1}{2}bh$$

$$A = \frac{1}{2}(16 - 2x)(4\sqrt{(x - 4)})$$

$$A = (8 - x)(2\sqrt{(x - 4)})$$

$$A = (8 - x)[2(x - 4)^{-1/2}]$$

$$\frac{dA}{dx} = \frac{1}{2}(2)(x - 4)^{-1/2}(8 - x) + (2)(x - 4)^{1/2}(-1)$$

$$\frac{dA}{dx} = (x - 4)^{-1/2}(8 - x) - (2)(x - 4)^{1/2}$$

$$\frac{dA}{dx} = \frac{(8 - x) - (2)(x - 4)}{\sqrt{x - 4}}$$

$$\frac{dA}{dx} = \frac{8 - x - 2x + 8}{\sqrt{x - 4}}$$

$$\frac{dA}{dx} = \frac{16 - 3x}{\sqrt{x - 4}}$$

$$0 = \frac{16 - 3x}{\sqrt{x - 4}}$$

$x = 4$ leads to an undefined solution

$$0 = 16 - 3x$$

$$3x = 16$$

$$x = \frac{16}{3} = 5.33 \text{ cm}$$

$$(16 - 2x) = \left[16 - (2)\left(\frac{16}{3}\right)\right] = \left[16 - \frac{32}{3}\right] = \left[\frac{48}{3} - \frac{32}{3}\right] = \frac{16}{3} = 5.33 \text{ cm}$$

Therefore, the triangle is an equilateral triangle.

(C) What is the maximum area?

$$h = 4\sqrt{(x - 4)} = 4\sqrt{\left(\frac{16}{3} - 4\right)} = 4\sqrt{\left(\frac{16}{3} - \frac{12}{3}\right)} = 4\sqrt{\frac{4}{3}} = 1.16 \text{ cm}$$

$$A = \frac{1}{2}bh$$

$$A = \frac{1}{2}\left(\frac{16}{3}\right)\left(4\sqrt{\frac{4}{3}}\right)$$

$$A = \left[(2)\left(\frac{16}{3}\right)\sqrt{\frac{4}{3}}\right]$$

$$A = \left[\frac{32}{3}\sqrt{\frac{4}{3}}\right]$$

$$A = (10.67 \text{ cm})(1.16 \text{ cm})$$

$$A = 12.3 \text{ cm}^2$$

199. A T-shirt maker estimates that the weekly cost of making x shirts is $C(x) = 50 + 2x + x^2/20$. The weekly revenue from selling x shirts is given by the function: $R(x) = 20x + x^2/200$ (show your work).

(A) What is the profit if all the shirts made are sold?

$$P(x) = R(x) - C(x)$$

$$R(x) = 20x + \frac{x^2}{200}$$

$$C(x) = 50 + 2x + \frac{x^2}{20}$$

$$P(x) = 20x + \frac{x^2}{200} - \left(50 + 2x + \frac{x^2}{20}\right)$$

$$P(x) = 20x + \frac{x^2}{200} - 50 - 2x - \frac{x^2}{20}$$

$$P(x) = 18x - 50 - \frac{9x^2}{200}$$

(B) What is the maximum weekly profit?

$$P(x) = 18x - 50 - \frac{9x^2}{200}$$

$$P'(x) = 18 - (2)\left(\frac{9x}{200}\right)$$

$$P'(x) = 18 - 0.09x$$

$$0 = 18 - 0.09x$$

$$0.09x = 18$$

$$x = 200 \text{ shirts}$$

$$P(200) = 18(200) - 50 - \frac{9(200)^2}{200}$$
$$P(200) = 3600 - 50 - 1800$$
$$P(200) = \$1750$$

200. You are to make a cylindrical tin can with closed top to hold 360 cm³.

(A) Make a drawing and label what is given.

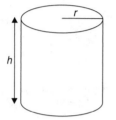

Variables:

- r = radius
- h = height
- V = volume
- S = surface area

Given: $V = 360$ cm³

(B) What are its dimensions if the amount of tin used is to be minimum?

$$V = \pi r^2 h$$
$$360 = \pi r^2 h$$
$$h = \frac{360}{\pi r^2}$$

$$S = 2\pi r h + 2\pi r^2$$
$$S = 2\pi r \left(\frac{360}{\pi r^2} \right) + 2\pi r^2$$
$$S = 2 \left(\frac{360}{r} \right) + 2\pi r^2$$
$$S = 720r^{-1} + 2\pi r^2$$
$$\frac{dS}{dr} = -720r^{-2} + 4\pi r$$
$$0 = \frac{-720}{r^2} + 4\pi r$$
$$4\pi r = \frac{720}{r^2}$$

$$4\pi r^3 = 720$$

$$r^3 = \frac{720}{4\pi}$$

$$r^3 = \frac{180}{\pi}$$

$$r = \sqrt[3]{\frac{180}{\pi}}$$

$$r = 3.86 \text{ cm}$$

$$h = \frac{360}{\pi r^2} = \frac{360}{\pi (3.86 \text{ cm})^2} = 7.71 \text{ cm}$$

(C) What is the surface area?

$$S = 2\pi rh + 2\pi r^2$$

$$S = 2\pi(3.86 \text{ cm})(7.71 \text{ cm}) + 2\pi(3.86 \text{ cm})^2$$

$$S = 187 \text{ cm}^2 + 94 \text{ cm}^2$$

$$S = 281 \text{ cm}^2$$

Chapter 5: More Applications of Derivatives

201. (B) Two tangents are parallel if their slopes, or $\dfrac{dy}{dx}$ of the function at each point, are equal.

202. (B) The instantaneous velocity $|v| = (3)^3 - 12(3)^2 + 5 = -76$.

203. (E) To make a linear approximation, let $f(x) = x^{1/4}$, $a = 81$, and $\Delta x = -3$.

$$f(a + \Delta x) \approx f(81) + f'(81)(-3) \approx \sqrt[4]{81} + \frac{1}{4}(81)^{-3/4}(-3) \approx 3 - \frac{1}{4}\left(\frac{3}{27}\right) = 3 - \frac{1}{36} \approx 2.97$$

204. (C) $\dfrac{dr}{d\theta} = \cos\theta = \cos\dfrac{\pi}{3} = \dfrac{1}{2}$.

$$\frac{dy}{dx} = \frac{\left(\dfrac{dr}{d\theta}\right)\sin\theta + r\cos\theta}{\left(\dfrac{dr}{d\theta}\right)\cos\theta - r\sin\theta} = \frac{\left(\dfrac{1}{2}\right)\left(\dfrac{\sqrt{3}}{2}\right) + 2\left(\dfrac{1}{2}\right)}{\left(\dfrac{1}{2}\right)\left(\dfrac{1}{2}\right) - 2\left(\dfrac{\sqrt{3}}{2}\right)} = \frac{1 + \sqrt{3}}{1 - \sqrt{3}}$$

205. (C) $a = v' = \left\langle \dfrac{d^2x}{dt^2}, \dfrac{d^2y}{dt^2} \right\rangle = \langle 6t, -9 \rangle$. The magnitude of the acceleration, then, is

$$|a| = \sqrt{\frac{d^2x}{dt^2} + \frac{d^2y}{dt^2}} = \sqrt{324 + 81} \approx 20.1.$$

206. (B) $s(t) = 0$ when the penny hits the ground. So, $0 = -16t^2 + 200 \Rightarrow t = \dfrac{10\sqrt{2}}{4} \approx 4$ s.

207. (A) $m_{normal} = \dfrac{-1}{m_{tangent}} = -3$

208. (C) Using tangent line approximation and $f(5) = 10$, the point of tangency is $(5, 10)$. $f'(5) = 2$ so the slope of the tangent at $x = 5$ is $m_{tangent} = 2$. The equation for the tangent line is $y - 10 = 2(x - 5)$ or $y = 2x$. Therefore, $f'(5.5) \approx 2(5.5) \approx 11$.

209. (D) $\dfrac{dx}{dt} = 4t$ and $\dfrac{dy}{dt} = 4$. At $t = 1$, $\dfrac{dx}{dt} = 4$, $\dfrac{dy}{dt} = 4$. The speed $= \sqrt{\left(\dfrac{dx}{dt}\right)^2 + \left(\dfrac{dy}{dt}\right)^2} = \sqrt{16 + 16} = \sqrt{32} = 4\sqrt{2}$.

210. (B) $v(t) = s'(t) = 3t^2 - 18t + 24$. Set $v(t) = 0$ to determine when the particle undergoes a change of direction. $v(t) = 3(t^2 - 6t + 8) = 0$. $3(t - 4)(t - 2) = 0 \Rightarrow t = 2, 4$. At $t = 2, 4$ the particle changes direction, therefore it is moving to the left at $2 < t < 4$.

211. (B) Express the angle measurement in radians, $60° = \dfrac{\pi}{3}$. Let $f(x) = \cos x$, the approximation $a = \dfrac{\pi}{3}$ and $\Delta x = 2° = 2\left(\dfrac{\pi}{180}\right) = \dfrac{\pi}{90}$. $f'(x) = -\sin x$ and $f'\left(\dfrac{\pi}{3}\right) = -\sin\dfrac{\pi}{3} = -\dfrac{\sqrt{3}}{2}$. Now use linear approximation $f\left(\dfrac{\pi}{3} + \dfrac{\pi}{90}\right) \approx f\left(\dfrac{\pi}{3}\right) + f'\left(\dfrac{\pi}{3}\right)\left(\dfrac{\pi}{90}\right) \approx \cos\left(\dfrac{\pi}{3}\right) - \left(\sin\dfrac{\pi}{3}\right)\left(\dfrac{\pi}{90}\right) \approx \dfrac{1}{2} + \left(\dfrac{-\sqrt{3}}{2}\right)\left(\dfrac{\pi}{90}\right) \approx 0.5 - 0.03 = 0.47$.

212. (C) If $\dfrac{dy}{dx}\big|_{x=a} = 0$, then the function has horizontal tangents. If $\dfrac{dy}{dx}$ does not exist at $x = a$ but $\dfrac{dx}{dy}\big|_{x=a} w = 0$, then the function has vertical tangents at $x = a$.

213. (C) Let $f(x) = x^3$, $a = 5$, and $\Delta x = 0.2$. $f(a + \Delta x) \approx f(a) + f'(a)\Delta x$. Then $f(5 + .2) \approx f(5) + f'(5)(.2) = 125 + 3(5)^2(.2) = 125 + 15 = 140$.

214. (B) $a(t) = v'(t) = 6t^2 - t + 4$. At $t = 4$, $a(4) = 96$ ft/s^2.

215. (C) Differentiate $\dfrac{dy}{dt} = -4t + 5$ and $\dfrac{dx}{dt} = 6$. When $t = 1$, $\dfrac{dy}{dt} = 1$. The speed $s = \sqrt{\left(\dfrac{dy}{dt}\right)^2 + \left(\dfrac{dx}{dt}\right)^2} = \sqrt{1^2 + 6^2} = \sqrt{37}$.

216. (A) Differentiate $\dfrac{dy}{dt} = \dfrac{2}{3}t$ and $\dfrac{dx}{dt} = 6$. Thus, $\dfrac{dy}{dt}\big|_{t=3} = 2$ and $\dfrac{dx}{dt}\big|_{t=3} = 6$. The slope of the tangent, $m_{\text{tangent}} = \dfrac{dy}{dt} \div \dfrac{dx}{dt} = \dfrac{2}{6} = \dfrac{1}{3}$.

217. (B) $T = \dfrac{r'(t)}{\|r'(t)\|} = \dfrac{\langle 3t^2,\, 4t \rangle}{\sqrt{144+64}} = \left\langle \dfrac{3t^2}{4\sqrt{13}},\, \dfrac{t}{\sqrt{13}} \right\rangle = \left\langle \dfrac{3}{\sqrt{13}},\, \dfrac{2}{\sqrt{13}} \right\rangle$

218. (B) $\dfrac{dx}{dt} = 2\cos t$ and $\dfrac{dy}{dt} = 2t - 2 = 2(t-1)$. Thus, $\dfrac{dy}{dx} = \dfrac{t-1}{\cos t}$.

219. (D) The acceleration vector $a = v' = \langle 8t,\, 3 \rangle$. The magnitude of the acceleration $|a| = \sqrt{256+9} = \sqrt{265} = 16.3$.

220. (B) Take the derivative of each function of y and set them equal to each other. So, $\dfrac{dy}{dx} = 2x = 4 \Rightarrow x = 2$.

221. (E) $\dfrac{dr}{d\theta} = -2\sin\theta = -\sqrt{3}$ when $\theta = \dfrac{\pi}{3}$.

$$\dfrac{dy}{dx} = \dfrac{\left(\dfrac{dr}{d\theta}\right)\sin\theta + r\cos\theta}{\left(\dfrac{dr}{d\theta}\right)\cos\theta - r\sin\theta} \text{ by the chain rule. So } \dfrac{dy}{dx} = \dfrac{(-\sqrt{3})\left(\dfrac{\sqrt{3}}{2}\right) + 2\left(\dfrac{1}{2}\right)}{(-\sqrt{3})\left(\dfrac{1}{2}\right) - 2\left(\dfrac{\sqrt{3}}{2}\right)} = \dfrac{1}{3\sqrt{3}}.$$

222. (E) $v_0 = 0$ since the ball starts from a rest position. $s(t) = 0$ when the ball hits the ground. Now we have $s(t) = -16t^2 + s_0$. Thus when $t = 3$, $s_0 = 16t^2 = 16(15)^2 = 3600$ ft.

223. (C) $\dfrac{dx}{dt} = -9t^2 - 6t + 6$ and $\dfrac{dy}{dt} = 3\cos t$. $\dfrac{dy}{dx} = \dfrac{dy/dt}{dx/dt} = \dfrac{3\cos t}{-9t^2 - 6t + 6} = \dfrac{\cos t}{2 - 2t - 3t^2}$.

224. (A) First, find $v(t) = s'(t) = 3t^2 + 2t - 5$. A change of direction occurs when $v(t) = 0$. $(3t + 5)(t - 1) = 0 \Rightarrow t = 1, -\dfrac{5}{3}$ but $t \geq 0$ so $t = 1$. To check that a change of direction actually occurs, you can look at $s(0) = 1$, $s(1) = -2$, $s(2) = 8$. Since there is a change of sign in the position of $s(t)$, you can infer a change of direction.

225. (A) Using tangent line approximation, you have $f(3) = \dfrac{1}{3} \Rightarrow$ the point of tangency is $\left(3, \dfrac{1}{3}\right)$. $f'(3) = -\dfrac{1}{9} \Rightarrow$ the slope of the tangent at $x = 3$ is $-\dfrac{1}{9}$. The equation of the tangent line is $y - 3 = -\dfrac{1}{9}(x - 3)$ or $y = -\dfrac{1}{9}x + \dfrac{2}{3}$. Inserting 3.4 for x yields $y = -\dfrac{1}{9}(3.4) + \dfrac{2}{3} = 0.288$.

226. (C) Find $\dfrac{dy}{dx}$ for each function of y. $\dfrac{dy}{dx} = 6x^2 \cdot \dfrac{dy}{dx} = x + 1$. If the tangents are parallel, their derivatives are equal, so set $6x^2 = x + 1 \Rightarrow (3x + 1)(2x - 1) = 0 \Rightarrow x = -\dfrac{1}{3}, \dfrac{1}{2}$.

227. (C) Using $f(a + \Delta x) \approx f(a) + f'(a)\Delta x$, set $f(x) = \sqrt{x}$, $a = 144$, $\Delta x = -5$. Then $f'(x) = -\dfrac{1}{2}x^{-1/2}$. Thus, $f(144 - 5) \approx f(144) - f'(144)(5) \approx \sqrt{144} + \dfrac{1}{2}\left(\dfrac{-5}{\sqrt{144}}\right) = 12 - \dfrac{5}{24} = 11.79$.

228. (D) Differentiate $\dfrac{dx}{dt} = -8t + 2$ and $\dfrac{d^2x}{dt^2} = -8$. Differentiate $\dfrac{dy}{dt} = 6$ and $\dfrac{d^2y}{dt^2} = 0$. Therefore the acceleration vector is $\langle -8, 0 \rangle$.

229. (D) $s = \sqrt{\left(\dfrac{dx}{dt}\right)^2 + \left(\dfrac{dy}{dt}\right)^2} = \sqrt{(3t^2)^2 + t^2} = t\sqrt{9t^2 + 1}$

230. (A) The velocity $v = r' = \left\langle \dfrac{1}{2}t^{-1/2}, 3t^2 \right\rangle$. The acceleration $a = v' = \left\langle -\dfrac{1}{4}t^{-3/2}, 6t \right\rangle$.

Evaluated at $t = 4$, the magnitude of the acceleration $|a| = \sqrt{\left[-\dfrac{1}{4}(4)^{-3/2}\right]^2 + [6(4)]^2} = \sqrt{\dfrac{1}{16}(4)^{-3} + 576} \approx \sqrt{576} = 24$.

231. (B) $a(t) = v'(t) = 2t - 2$. Over the interval $0 \le t \le 5$, $a(t)$ is initially at $a(0) = -2$, which represents a deceleration, and increases to $a(5) = 8$. The increase in $a(t) = 8 - (-2) = 10$.

232. (C) The slope of the tangent to $y = x^2 - 2$ is $\dfrac{dy}{dx} = 2x$. The slope of $y = \dfrac{1}{2}x + a$ is $\dfrac{1}{2}$.

Thus, $\dfrac{dy}{dx} = 2x = \dfrac{1}{2}$ for $x = \dfrac{1}{4}$. At $x = \dfrac{1}{4}$, $y = \left(\dfrac{1}{4}\right)^2 - 2 = \dfrac{1}{16} - \dfrac{32}{16} = \dfrac{-31}{16} \Rightarrow \left(\dfrac{1}{4}, \dfrac{-31}{16}\right)$ is a

tangent point. So $y = \dfrac{1}{2}x + a \Rightarrow \dfrac{-31}{16} = 3\left(\dfrac{1}{4}\right) + a \Rightarrow a = -\dfrac{43}{16}$.

233. (A) When $t = 0$, $\langle e^t, e^{-t} \rangle = (1, 1)$. $\dfrac{dy}{dx} = -\dfrac{e^{-t}}{e^t}\Big|_{t=0} = -1$ so the equation of the tangent line is $(y - 1) = -1(x - 1) \Rightarrow y = 2 - x$.

234. (B) Apply the product rule to find $\dfrac{dx}{d\theta} = r\sec^2\theta + \tan\theta\dfrac{dr}{d\theta}$ and $\dfrac{dy}{d\theta} = r\sec\theta\tan\theta + \sec\theta\dfrac{dr}{d\theta}$.

Therefore, $\dfrac{dy}{dx} = \dfrac{\dfrac{dy}{d\theta}}{\dfrac{dx}{d\theta}} = \dfrac{r\sec\theta\tan\theta + \sec\theta\dfrac{dr}{d\theta}}{r\sec^2\theta + \tan\theta\dfrac{dr}{d\theta}}$.

235. (B) When the ball hits the ground, $s(t) = 0 \Rightarrow -16t^2 + 40 = 0 \Rightarrow t = \pm\dfrac{\sqrt{10}}{2}$, but $t \geq 0$ so $t = \dfrac{\sqrt{10}}{2}$.

236. (B) First, express the angle measurement in radians. $120° = \dfrac{2\pi}{3}$ radians and $2° = \dfrac{\pi}{90}$ radians. Let $f(x) = \sin x$, a = the approximation = $\dfrac{2\pi}{3}$ and $\Delta x = \dfrac{\pi}{90}$. $f'(x) = \cos x$ and $f'\left(\dfrac{2\pi}{3}\right) = -\dfrac{1}{2}$. $f\left(\dfrac{2\pi}{3} + \dfrac{\pi}{90}\right) \approx f\left(\dfrac{2\pi}{3}\right) + f'^{\left(\frac{2\pi}{3}\right)\left(\frac{\pi}{90}\right)} \approx \sin\left(\dfrac{2\pi}{3}\right) + \left[\cos\left(\dfrac{2\pi}{3}\right)\right]\left(\dfrac{\pi}{90}\right) \approx \dfrac{\sqrt{3}}{2} + \left(-\dfrac{1}{2}\right)\left(\dfrac{\pi}{90}\right) = 0.849$

237. (C) $v(t) = s'(t) = 9t^2 - 25$. At time $t = 2$, $v(2) = 32$ ft/s.

238. (A) $f'(x) = 3x^2 + 2$ and $f'(-1) = 5$. An equation for the tangent is then $y = f(-1) + f'(-1)(x+1) = (-1)^3 + 2(-1) - 1 + 5(x+1) = -1 - 3 + 5(x+1) \Rightarrow y = 5x + 1$.

239. (C) $\dfrac{dy}{dx} = \dfrac{\dfrac{dy}{dt}}{\dfrac{dx}{dt}} = \dfrac{3t^2 - 6t}{-3\sin t} = -\dfrac{t(t-2)}{\sin t}$

240. (B) For horizontal tangents, $\dfrac{dy}{dx} = 0$. Since $\dfrac{dy}{dx} = 2x$, this occurs at $x = 0$. Substituting back into the original equation for the curve, when $x = 0$, $y = -1$.

241. (E) $\dfrac{dx}{dt} = 4t$ and $\dfrac{dy}{dt} = -5$. When $t = 2$, $\dfrac{dx}{dt} = 8$, $\dfrac{dy}{dt} = -5$. Then speed $= \sqrt{8^2 + (-5)^2}$ $= \sqrt{64 + 25} = \sqrt{89}$.

242. (B) $v(t) = s'(t) = 10t - 12$. $a(t) = v'(t) = 10$. At $t = 2$, $v(2) = 8$ ft/s and $a(2) = 10$ ft/s^2.

243. (B) The slope of the tangent, $m_{\text{tangent}} = \dfrac{dy}{dx}\Big|_{x=\frac{\pi}{4}} = -6\sin 2\left(\dfrac{\pi}{4}\right) = -6$. The slope of the normal line $m_{\text{normal}} = -\dfrac{1}{m_{\text{tangent}}} = \dfrac{1}{6}$.

244. (C) $f(4) = 48 \Rightarrow$ point of tangency is $(4, 48)$. $f'(4) = 24 \Rightarrow$ the slope of the tangent at $x = 4$ is $m_{\text{tangent}} = 24$. The equation of the tangent is $y - 48 = 24(x - 4)$ or $y = 24(x - 2)$. Therefore, $f(4.3) = 24(4.3 - 2) = 24(2.3) = 55.2$.

245. (B) $v(t) = 9t^2 - 6t$ and $a(t) = v'(t) = 18t - 6$ so the acceleration equals 0 when $t = \dfrac{1}{3}$.

246. (A) The slope of the tangent, $m_{tangent}$, is defined as $\dfrac{dy}{dx} \cdot \dfrac{dy}{dx} = 3x^2$. At $x = 2$, $\dfrac{dy}{dx} = 12$.

(B) $m_{normal} = -\dfrac{1}{m_{tangent}} = -\dfrac{1}{3x^2} = -\dfrac{1}{12}$ at $x = 2$.

(C) Since the slope of the normal line is $-\dfrac{1}{12}$, and, taking the point $(2, 4)$, we can write $\dfrac{y-4}{x-2} = -\dfrac{1}{12}$ or $y = -\dfrac{1}{12}x + 4\dfrac{1}{6}$.

247. (A) $v(t) = s'(t) = -32t + 256$

(B) At the maximum altitude, $v(t) = 0$ so $-32t + 256 = 0$ or $t = 8$ s. The projectile reaches its maximum altitude in 8 s. Find its position at 8 s by applying $s(8) = -16(8)^2 + 256(8) = -1024 + 2048 = 1024$ ft.

(C) $a(t) = v'(t) = -32$ ft/s^2

(D) When the projectile hits the ground, $s(t) = 0$. Therefore $-16t^2 + 256t = 0$. Factoring to solve for t, we find that $t = 0, 16$. The projectile lands in 16 s.

248. (A) $v(t) = r'(t) = \langle -2\sin t, \ 2\cos t \rangle$. At $\dfrac{\pi}{4}$, $v(t) = \left\langle -2\left(\dfrac{\sqrt{2}}{2}\right), \ 2\left(\dfrac{\sqrt{2}}{2}\right) \right\rangle = \langle -\sqrt{2}, \ \sqrt{2} \rangle = \langle -1.414, \ 1.414 \rangle$.

(B) $a(t) = v'(t) = \langle -2\cos t, \ -2\sin t \rangle$. At $\dfrac{\pi}{4}$,

$a(t) = \left\langle -2\left(\dfrac{\sqrt{2}}{2}\right), \ -2\left(\dfrac{\sqrt{2}}{2}\right) \right\rangle = \langle -\sqrt{2}, \ -\sqrt{2} \rangle = \langle -1.414, \ -1.414 \rangle$.

(C) $T(t) = \dfrac{r'(t)}{\| r'(t) \|} = \dfrac{\langle -2\sin t, \ 2\cos t \rangle}{\sqrt{(-2\sin t)^2 + (2\cos t)^2}}$ which simplifies to $\langle -\sin t, \ \cos t \rangle$. At

time $\dfrac{\pi}{4}$, $T(t) = \left\langle \dfrac{-\sqrt{2}}{2}, \ \dfrac{\sqrt{2}}{2} \right\rangle = \langle -0.707, \ 0.707 \rangle$.

(D) $N(t) = \dfrac{T'(t)}{\| T'(t) \|} = \dfrac{\langle -\cos t, \ -\sin t \rangle}{\sqrt{(-\cos t)^2 + (-\sin t)^2}}$ which simplifies to $\langle -\cos t, \ -\sin t \rangle$.

At time $\dfrac{\pi}{4}$, $N(t) = \left\langle -\dfrac{\sqrt{2}}{2}, \ -\dfrac{\sqrt{2}}{2} \right\rangle = \langle -0.707, \ -0.707 \rangle$.

249. (A) The slope of the tangent line equals $\dfrac{dy}{dx} = \dfrac{dy/dt}{dx/dt} = \dfrac{2t-5}{3t^2-2} = -3$ at $t = -1$.

Substituting $t = 1$ into the parametric equations gives us the coordinates $(0, -4)$. Thus, the equation for the tangent line is $y + 4 = -3(x - 0)$ or, solving for y, $y = -3x - 4$.

(B) The tangent line is horizontal if $\dfrac{dy}{dt} = 0$. Thus $2t - 5 = 0$ and $t = \dfrac{5}{2}$.

(C) The tangent line is vertical if $\dfrac{dx}{dt} = 0$. Thus $3t^2 - 2 = 0$ and $= \pm\sqrt{\dfrac{2}{3}}$.

250. (A) $v(t) = s'(t) = 32t$. $v(1) = -32$ ft/s

(B) $v_{\text{average}} = \dfrac{s(2) - s(0)}{2} = \dfrac{-16(2)^2 + 150 - 150}{2} = -32$ ft/s

(C) $s(t) = 0$ when the penny hits the ground so $-16t^2 + 150 = 0 \Rightarrow t = \pm\sqrt{\dfrac{150}{16}} =$

$\pm\dfrac{5\sqrt{6}}{4}$. Since you also know that $t \geq 0$, $t = \dfrac{5\sqrt{6}}{4} = 3.06$ s.

Chapter 6: Integration

251. (D) Apply the formula $\displaystyle\int x^n = \dfrac{x^{n+1}}{n+1} + C$. Thus $\displaystyle\int (x^4 - 3x^2 + 1)dx = \dfrac{x^5}{5} - x^3 + x + C$.

252. (B) Integrate using partial fractions. Let A and B represent the numerators of the

partial fractions $\dfrac{1}{x(x+2)} = \dfrac{A}{x} + \dfrac{B}{x+2}$. Then $A(x+2) + Bx = 1 \Rightarrow Ax + Bx = 0$ and $2A = 1$.

Therefore, $A = \dfrac{1}{2}$, $B = -\dfrac{1}{2}$. Substituting into the integral give us, $\displaystyle\int \dfrac{1}{x(x+2)} = \int \dfrac{1}{2x}dx -$

$\displaystyle\int \dfrac{1}{2(x+2)}dx = \dfrac{1}{2}\ln|x| - \dfrac{1}{2}\ln|x+2| + C$.

253. (A) Let $u = x^3$. Differentiate $du = 3x^2\,dx \Rightarrow \dfrac{du}{3} = x^2\,dx$. Rewrite $\displaystyle\int \dfrac{1}{6}\sin u\,du$. Integrate:

$-\dfrac{1}{6}\cos u + C$. Replace u: $-\dfrac{1}{6}\cos x^3 + C$.

254. (E) Rewrite $\displaystyle\int \dfrac{1}{x^2 + 6x + 9 + 4}dx = \int \dfrac{1}{(x+3)^2 + 2^2}dx$. Let $u = x + 3$, $du = dx$. Rewrite

$\displaystyle\int \dfrac{1}{u^2 + 2^2}du = \dfrac{1}{2}\tan^{-1}\left(\dfrac{u}{2}\right) + C$. Replace u: $\dfrac{1}{2}\tan^{-1}\left(\dfrac{x+3}{2}\right) + C$.

255. (C) Factor the numerator to simplify: $\displaystyle\int \dfrac{(x+5)(x-2)}{(x-2)}dx = \int (x+5)dx$. Integrate:

$\dfrac{1}{2}x^2 + 5x + C$.

256. (B) Let $u = x^2 + 1$, $du = 2x\,dx \Rightarrow \dfrac{du}{2} = x\,dx$. Rewrite $\displaystyle\int e^u\,du$. Integrate: $e^u + C$.

Replace u: $e^{x^2+1} + C$.

257. (D) Let $u = 3x$, $dv = \cos 3x\,dx$. Thus $du = 3\,dx$, $v = \dfrac{1}{3}\sin 3x$. Then $\displaystyle\int 3x\cos 3x\,dx =$

$x\sin 3x - \displaystyle\int \dfrac{1}{3}\sin 3x(3) = x\sin 3x + \dfrac{1}{3}\cos 3x + C$.

258. (A) Factor the denominator $\int \dfrac{dx}{(x-6)(x+1)}$. Let A and B represent the numerators of

the partial fractions $\dfrac{1}{(x-6)(x+1)} = \dfrac{A}{x-6} + \dfrac{B}{x+1}$. $A(x+1) + B(x-6) = 1$ so $Ax + Bx = 0$.

Solve for A and B so that $A = \dfrac{1}{7}$, $B = -\dfrac{1}{7}$. Thus $\int \dfrac{dx}{x^2 - 5x - 6} = \int \dfrac{1/7}{(x-6)} dx + \int \dfrac{-1/7}{(x+1)} dx = $

$\dfrac{1}{7} \ln |x-6| - \dfrac{1}{7} \ln |x+1| + C.$

259. (E) Use integration by parts. Let $u = e^x$, $dv = \sin x\, dx$. Thus $du = e^x dx$ and $v = -\cos x$.
Then $\int e^x \sin x\, dx = -e^x \cos x + \int e^x \cos x + C$. Substitute a second time to solve the new
integral. Let $u = e^x$ and $dv = \cos x\, dx$. Then $du = e^x dx$ and $v = \sin x$. Now we find that
$\int e^x \sin x\, dx = -e^x \cos x + e^x \sin x - \int e^x \sin x\, dx + C$. Combining terms, we find that
$2\int e^x \sin x\, dx = e^x(\sin x - \cos x) + C$ or $\int e^x \sin x = \dfrac{1}{2} e^x(\sin x - \cos x) + C.$

260. (C) Rewrite as $\int \left(5 - \dfrac{2}{x} + \dfrac{1}{x^2} \right) dx = 5x - 2 \ln |x| + \dfrac{1}{x} + C.$

261. (A) $\int \sec x \tan x\, dx = \sec x + C$ so that $\int (5 \sec x)(2 \tan x) dx = 10 \sec x + C.$

262. (B) Apply the formula $\int a^x dx = \dfrac{a^x}{\ln a} + C$, $a > 0$, $a \neq 1$. Thus $\int 5^x dx = \dfrac{5^x}{\ln 5} + C.$

263. (C) $4\int (\ln(x) + \sin^2 x) dx = 4\int \ln(x) dx + 4\int \sin^2 x\, dx = 4x \ln |x| - 4x + 4\left(\dfrac{x}{2} \right) - $

$4\left[\dfrac{\sin 2x}{4} \right] + C = 4x \ln |x| - 2x - \sin 2x + C.$

264. (D) Let $u = x^3 + 1$, then $du = 3x^2 dx \Rightarrow \dfrac{du}{3} = x^2 dx$. Rewrite and integrate:

$\int \dfrac{du}{3x^3} = -\dfrac{1}{6} u^{-2} + C$. Replace u: $-\dfrac{1}{6(x^3 + 1)^2} + C.$

265. (A) $\int \dfrac{\sin x - \cos x}{\cos x} dx = \int \dfrac{\sin x}{\cos x} dx - \int \dfrac{\cos x}{\cos x} dx = \int \tan x\, dx - \int dx = \ln |\sec x| - x + C.$

266. (D) Let $u = 3x$ and $dv = \sec^2 x\, dx$. Thus $du = 3\, dx$ and $v = \tan x + C$. Rewrite:
$3x \tan x - \int 3 \tan x\, dx = 3x \tan x - 3 \ln |\sec x| + C.$

267. (E) $\int \dfrac{2x+3}{x} dx = \int \dfrac{2x}{x} dx + \int \dfrac{3}{x} dx = \int 2\, dx + \int \dfrac{3}{x} dx = 2x + 3 \ln |x| + C.$

268. (A) Let $u = 3x$, $du = 3\, dx \Rightarrow \dfrac{du}{3} = dx$. Rewrite: $\dfrac{1}{3} \int 7^u du$. Integrate: $\dfrac{1}{3} \dfrac{7^u}{\ln(7)} + C = $

$\dfrac{7^{3x}}{3\ln(7)} + C.$

269. (C) Factor the denominator: $\int \dfrac{dx}{(x+2)(x-2)}$. Use partial fractions to solve

$\dfrac{1}{(x+2)(x-2)} = \dfrac{A}{x-2} + \dfrac{B}{x+2}$. $A(x+2) + B(x-2) = 1 \Rightarrow Ax + Bx = 0$, $2A - 2B = 1$.

Solve for A and B to find $A = \dfrac{1}{4}$, $B = -\dfrac{1}{4}$. Therefore $\int \dfrac{dx}{x^2-4} = \int \dfrac{1/4}{x-2}dx + \int \dfrac{-1/4}{x+2}dx = $

$\dfrac{1}{4}\ln|x-2| - \dfrac{1}{4}\ln|x+2| + C$.

270. (E) $\dfrac{1}{2}\int 4^x \ln(4)dx = \dfrac{1}{2}\ln(4)\int 4^x\, dx = \dfrac{4^x \ln(4)}{2\ln(4)} + C = \dfrac{1}{2}4^x + C$

271. (D) $\int (x^2 - \sin^2 x)dx = \int x^2 dx - \int \sin^2 x\, dx = \dfrac{1}{3}x^3 - \int \left(\dfrac{1-\cos 2x}{2}\right)dx = \dfrac{1}{3}x^3 - $

$\int \dfrac{1}{2}dx + \int \dfrac{\cos 2x}{2}dx = \dfrac{1}{3}x^3 - \dfrac{1}{2}x + \dfrac{1}{4}\sin 2x + C$

272. (C) Apply the formula $\int \dfrac{1}{x\sqrt{x^2 - a^2}}dx = \dfrac{1}{a}\sec^{-1}\left|\dfrac{x}{a}\right| + C$. Then $\int \dfrac{4}{x\sqrt{x^2 - 4}}dx = $

$4\left(\dfrac{1}{2}\right)\sec^{-1}\left|\dfrac{x}{2}\right| + C = 2\sec^{-1}\left|\dfrac{x}{2}\right| + C$.

273. (A) Let $u = x - 3$, so $x = u + 3$. Differentiate to get $du = dx$. Rewrite integral:

$\int (u+3)u^3\, du = \int (u^4 + 3u^3)du = \dfrac{u^5}{5} + \dfrac{3}{4}u^4 + C$. Replace u: $\dfrac{1}{5}(x-3)^5 + \dfrac{3}{4}(x-3)^4 + C$.

274. (B) Let $u = x^3 - 1$ then $du = 3x^2 dx$. Rewrite the integral: $\int e^u du = e^u + C$. Replace u: $e^{(x^3-1)} + C$.

275. (A) Let $u = x^3$, then $du = 3x^2 dx \Rightarrow \dfrac{du}{3} = x^2 dx$. Thus $\int \dfrac{x^2}{\sqrt{1-x^6}} = \dfrac{1}{3}\int \dfrac{3x^2}{\sqrt{1-(x^3)^2}} = $

$\dfrac{1}{3}\int \dfrac{du}{\sqrt{1-u^2}} = \dfrac{1}{3}\sin^{-1}(x^3) + C$.

276. (C) Use integration by parts. Let $u = x^2$, $dv = e^{-x}dx$. Thus $du = 2x\, dx$, $v = -e^{-x}$. Then $\int x^2 e^{-x}dx = -x^2 e^{-x} - \int (-e^{-x})2x\, dx$. Integrate by parts a second time. Let $u = 2x$, $du = 2\, dx$, $dv = e^{-x}dx$ and $v = -e^{-x}$. Then the integration yields $-x^2 e^{-x} - 2xe^{-x} - 2e^{-x} + C = -e^{-x}(x^2 + 2x + 2) + C$.

277. (E) Let $u = 3x$, $du = 3dx \Rightarrow \dfrac{du}{3}dx$. Rewrite $\int 9^{3x}dx = \dfrac{1}{3}\int 9^u du = \dfrac{1}{3}\dfrac{9^u}{\ln 9} + C$. Replace u to get $\dfrac{1}{3}\dfrac{9^{3x}}{\ln 9} + C$.

278. (D) $\int x^{1/2}(x^3+1)dx = \int (x^{7/2}+x^{1/2})dx = \dfrac{x^{9/2}}{\frac{9}{2}}+\dfrac{x^{3/2}}{\frac{3}{2}}+C = \dfrac{2}{9}x^{9/2}+\dfrac{2}{3}x^{3/2}+C.$ By

combining terms and simplifying, we find that the result is $\dfrac{2x^{9/2}+6x^{3/2}}{9}+C =$

$\dfrac{\sqrt{x}\,(2x^4+6x)}{9}+C.$

279. (C) Use integration by parts. Let $u = \sec x$, $dv = \sec^2 x\,dx$. Then $du = \sec x\,\tan x\,dx$ and $v = \tan x$. Now $\int \sec^3 x\,dx = \sec x\tan x - \int \tan^2 x\sec x\,dx.$ Apply the relation $\tan^2 x = \sec^2 x - 1$ to substitute for $\tan^2 x$ and find $\int \sec^3 x = \sec x\,\tan x - \int \sec x(\sec^2 x - 1)dx = \sec x\tan x - \int \sec^3 x\,dx + \int \sec x\,dx.$ Then $\int \sec^3 x = \sec x\tan x - \int \sec^3 x\,dx + \ln|\sec x+\tan x|+C.$ Combining terms, we obtain $2\int \sec^3 x\,dx = \sec x\tan x + \ln|\sec x + \tan x| + C.$ Then, dividing by two on both sides, $\int \sec^3 x\,dx = \dfrac{1}{2}(\sec x\tan x + \ln|\sec x + \tan x|) + C.$

280. (B) $\int \left(x^4 + \dfrac{x}{9+x^4}\right)dx = \int x^4 dx + \int \dfrac{x}{9+(x^2)^2}dx.$ Use substitution to solve the

second integral. Let $u = x^2$, $du = 2x\,dx \Rightarrow \dfrac{du}{2} = x\,dx.$ Then $\int \left(x^4 + \dfrac{x}{9+x^4}\right)dx = \dfrac{1}{5}x^5 +$

$\dfrac{1}{2}\int \dfrac{du}{(3^2+u^2)} = \dfrac{1}{5}x^5 + \dfrac{1}{2}\left(\dfrac{1}{3}\right)\tan^{-1}\left(\dfrac{u}{3}\right) + C.$ Substitute u: $\dfrac{1}{5}x^5 + \dfrac{1}{6}\tan^{-1}\left(\dfrac{x^2}{3}\right) + C.$

281. (C) $\int x(x-5)(x+2)dx = \int (x^3 - 3x^2 - 10x)dx = \dfrac{1}{4}x^4 - x^3 - 5x^2 + C$

282. (A) Let $u = x^{1/2}$, $du = \dfrac{1}{2}x^{-1/2}dx \Rightarrow 2du = x^{-1/2}dx.$ Then $\int \dfrac{\sin\sqrt{x}}{\sqrt{x}}dx = 2\int \sin u\,du =$ $-2\cos u + C.$ Substitute u: $-2\cos\sqrt{x} + C.$

283. (D) $\int \dfrac{3x^2+x-6}{x^2}\,dx = \int 3\,dx + \int \dfrac{1}{x}dx - \int 6x^{-2}dx = 3x + \ln|x| + \dfrac{6}{x} + C$

284. (E) Let $u = 1+x^2$, $du = 2x\,dx.$ Then $\int 2x\sqrt{1+x^2}\,dx = \int u^{1/2}du = \dfrac{2}{3}u^{3/2} + C.$ Substitute u: $\dfrac{2}{3}(1+x^2)^{3/2} + C.$

285. (D) Begin by factoring the denominator $\int \dfrac{3x+5}{x^2+3x-4}\,dx.$ Apply integra-

tion by partial fractions $\dfrac{3x+5}{(x+4)(x-1)} = \dfrac{A}{x+4} + \dfrac{B}{x-1}.$ $3x+5 = A(x-1) + B(x+4).$

Answers ❮ 209

Thus, $Ax + Bx = 3x$ and $-A + 4B = 5$. Solving, you find that $A = \dfrac{7}{5}$ and $B = \dfrac{8}{5}$.

$$\int \frac{3x+5}{x^2+3x-4}\,dx = \int \frac{7/5}{x+4}\,dx + \int \frac{8/5}{x+1}\,dx = \frac{7}{5}\ln|x+4| - \frac{8}{5}\ln|x-1| + C$$

286. (B) Let $u = 5x - 1$, $du = 5dx \Rightarrow \dfrac{du}{5} = dx$. Rewrite: $\displaystyle\int e^u\left(\frac{du}{5}\right) = \frac{1}{5}\int e^u\,du$. Integrate: $\dfrac{1}{5}e^u + C$. Replace u: $\dfrac{1}{5}e^{(5x-1)} + C$.

287. (C) Let $u = 3x$, $du = 3dx \Rightarrow \dfrac{du}{3} = dx$. Rewrite: $\displaystyle\int 21^u\left(\frac{du}{3}\right) = \frac{1}{3}\int 21^u\,du$. Integrate: $\dfrac{1}{3}\dfrac{21^u}{\ln(21)} + C = \dfrac{21^u}{3\ln(21)} + C$. Replace u: $\dfrac{21^{3x}}{3\ln(21)} + C$.

288. (D) Let $u = x - 1$ then $x = u + 1$ and $du = dx$. Rewrite: $\int (u+1)u^6\,du = \int (u^7 + u^6)\,du$. Integrate: $\dfrac{1}{8}u^8 + \dfrac{1}{7}u^7 + C$. Replace u: $\dfrac{1}{8}(x-1)^8 + \dfrac{1}{7}(x-1)^7 + C$.

289. (A) Rewrite the integral: $\int e^x\,dx - \int e^{3x}\,dx$. You can integrate the first integral as it stands, but you must use u-substitution on the second integral. Let $u = 3x$, $du = 3dx \Rightarrow \dfrac{du}{3} = dx$. You now have $\displaystyle\int e^x(1 - e^{2x})\,dx = e^x - \frac{1}{3}\int e^u\,du = e^x - \frac{1}{3}e^u + C = e^x - \frac{1}{3}e^{3x} + C$ after substituting back for u.

290. (B) Since $\ln x$ and e^x are inverse functions, the integral reduces to $\int (x^2 - x + 1)\,dx$. Integrate: $\dfrac{1}{3}x^3 - \dfrac{1}{2}x^2 + x + C$.

291. (C) Let $u = x^3 + 1$. Differentiate to find $du = 3x^2\,dx \Rightarrow \dfrac{du}{3} = x^2\,dx$. Rewrite: $\dfrac{1}{3}\displaystyle\int \frac{1}{u}\,du$. Integrate: $\dfrac{1}{3}\ln|u| + C$. Replace u: $\dfrac{1}{3}\ln|x^3 + 1| + C$.

292. (D) Use integration by parts where $u = \sin^3 x$ and $dv = \sin x\,dx$. Thus $du = 3\sin^2 x \cos x\,dx$ and $v = -\cos x$. Then $\int \sin^4 x\,dx = uv - \int v\,du = -\cos x \sin^3 x + \int 3\sin^2 x \cos^2 x\,dx$. Apply the relation $\cos^2 x = 1 - \sin^2 x$ to the new integral. Now $\int \sin^4 x = -\cos x \sin^3 x + 3\int \sin^2 x(1 - \sin^2 x)\,dx = -\cos x \sin^3 x + 3\int \sin^2 x\,dx - 3\int \sin^4 x\,dx$. Then combine like terms to get $4\int \sin^4 x = -\cos x \sin^3 x + 3\int \sin^2 x\,dx$. Apply the integration formula $\int \sin^2 x\,dx = \dfrac{x}{2} - \dfrac{\sin(2x)}{4}$. Now we have $4\int \sin^4 x\,dx = -\cos x \sin^3 x + \dfrac{3x}{2} - \dfrac{3\sin(2x)}{4} + C$. Then $\int \sin^4 x = -\dfrac{1}{4}\cos x \sin^3 x + \dfrac{3x}{8} - \dfrac{3\sin(2x)}{16} + C = -\dfrac{1}{4}\cos x \sin^3 x + \dfrac{3x}{8} - \dfrac{3}{8}\sin x \cos x + C$.

293. (A) Let $u = \ln x$, $du = \dfrac{1}{x}\,dx$. Rewrite in terms of u: $\dfrac{1}{4}\displaystyle\int u\,du$. Integrate: $\dfrac{1}{4}\left(\dfrac{1}{2}u^2\right) + C = \dfrac{1}{8}u^2 + C$. Replace u: $\dfrac{1}{8}(\ln x)^2 + C$.

294. (E) Rewrite $\displaystyle\int \frac{1}{x^2 - 6x + 9 + 4}\,dx = \int \frac{1}{(x-3)^2 + 2^2}\,dx$. Let $u = x - 3$, $du = dx$.

Rewrite: $\displaystyle\int \frac{1}{u^2 + 2^2}\,du$. Integrate: $\dfrac{1}{2}\tan^{-1}\left(\dfrac{u}{2}\right) + C$. Replace u: $\dfrac{1}{2}\tan^{-1}\left(\dfrac{x-3}{2}\right) + C$.

295. (C) Let $u = \cot x$ then $du = -\csc^2 x\,dx$. Substitute $\int \csc^2 x \, \cot x\,dx = -\int u\,du = -\dfrac{u^2}{2} + C = -\dfrac{1}{2}\cot^2 x + C$.

296. (A) At $x = 5$, $f(5) = \dfrac{2}{25 - 15 - 4} = \dfrac{2}{6} = \dfrac{1}{3}$ is the slope. Then the equation of the line

is $y - \dfrac{4}{5}\ln|6| = \dfrac{1}{3}(x - 5) \Rightarrow y = \dfrac{1}{3}x + \dfrac{4}{5}\ln|6| - \dfrac{5}{3}$.

(B) $f(4.5) \approx \dfrac{1}{3}(4.5) - \dfrac{5}{3} + \dfrac{4}{5}\ln 6 \approx 1.500 - 1.666 + 1.433 \approx 1.267$

(C) $f(x) = \displaystyle\int \frac{x-3}{x^2 - 3x - 4}\,dx$. Use integration by partial fractions. Factor

the denominator $\displaystyle\int \frac{x-3}{x^2 - 3x - 4}\,dx = \int \frac{x-3}{(x-4)(x+1)}\,dx$. Then $\dfrac{x-3}{(x-4)(x+1)} =$

$\dfrac{A}{x-4} + \dfrac{B}{x+1} \Rightarrow A(x+1) + B(x-4) = x - 3$. So, $Ax + Bx = x$, and $A - 4B = -3$.

Solving, we find that $A = \dfrac{1}{5}$ and $B = \dfrac{4}{5}$. Then $\displaystyle\int \frac{x-3}{x^2 - 3x - 4}\,dx = \int \frac{1/5}{x-4}\,dx +$

$\displaystyle\int \frac{4/5}{x+1}\,dx = \dfrac{1}{5}\ln|x-4| + \dfrac{4}{5}\ln|x+1| + C$. At $x = 5$, $f(5) = \dfrac{4}{5}\ln 6 \Rightarrow C = 0$.

(D) $f(4.5) = \dfrac{1}{5}\ln(4.5 - 4) + \dfrac{4}{5}\ln(4.5 + 1) \approx -0.139 + 1.364 \approx 1.225$

297. (A) $v(t) = \displaystyle\int (1 + t^{-1/2})\,dt = \int dt + \int t^{-1/2}\,dt = t + \dfrac{t^{1/2}}{\frac{1}{2}} + C = t + 2t^{1/2} + 2$

(B) $s(t) = \displaystyle\int (t + 2t^{1/2} + 2)\,dt = \dfrac{1}{2}t^2 + \dfrac{2t^{3/2}}{\frac{3}{2}} + 2t + C = \dfrac{1}{2}t^2 + \dfrac{4}{3}t^{3/2} + 2t + 10$

298. (A) $P(t) = \displaystyle\int (200 + 9t^{1/2} - 5t)\,dt = 200t + 9\dfrac{t^{3/2}}{\frac{3}{2}} - 5t^2 + C = 200t - 6t^{3/2} - \dfrac{5}{2}t^2 + C$

(B) $P(0) = 200(0) + 6(0) - \dfrac{5}{2}(0) + C \Rightarrow C = 2000$

(C) $P(4) = 200(4) + 6(4)^{3/2} - \dfrac{5}{2}(4)^2 + 2000 = 800 + 48 - 40 + 2000 = 2808$

299. (A) $MR_{\text{total}} = \displaystyle\int (100 - .5q)\,dq = 100q - \dfrac{1}{4}q^2 + C$. Since the manufacturer receives

nothing when nothing is produced, $MR_{\text{total}}(0) = 0 \Rightarrow C = 0$. Then $MR_{\text{total}} = 100q - \dfrac{1}{4}q^2$.

(B) $1000 = 100q - \frac{1}{4}q^2 \Rightarrow q^2 - 400q + 4000 = 0 \Rightarrow (q - 200)(q - 200) = 0$
$\Rightarrow q = 200$

(C) Revenue begins to decrease when $MR_{total} = 0$ or $100q = \frac{1}{4}q^2 \Rightarrow q = 400.$

300. **(A)** $m(x) = \int \frac{3}{2}\sqrt{x}\ dx = \frac{3}{2}\int x^{1/2}dx = \frac{3}{2}\frac{x^{3/2}}{\frac{3}{2}} + C = x^{3/2} + C.$ Since there is no mass

zero distance from the left side of the rod, $m(0) = 0.$ Thus, $C = 0.$

(B) $m(2) = (2)^{3/2} = \sqrt{8}$ g ≈ 2.83 kg

(C) When $x = 6$ meters, $m(x) = (6)^{3/2} = \sqrt{216}$ kg ≈ 14.70 kg.

Chapter 7: Definite Integrals

301. **(D)** $\sum_{i=1}^{5} i^3 = 1^3 + 2^3 + 3^3 + 4^3 + 5^3 = 1 + 8 + 27 + 64 + 125 = 225.$ Or use the formula

$\sum_{i=1}^{n} i^3 = \frac{n^2(n+1)^2}{4} = \frac{25(36)}{4} = 225.$

302. **(E)** $\int_{0}^{4} (x^3 - 3x + 1)\ dx = \frac{x^4}{4} - \frac{3x^2}{2} + x \Big]_{0}^{4} = (64 - 24 + 4) - 0 = 44$

303. **(A)** $\int_{\pi/2}^{\pi} \frac{1}{4 + x^2} dx = \frac{1}{2}\tan^{-1}\left(\frac{x}{2}\right)\Big]_{\pi/2}^{\pi} = \frac{1}{2}\tan^{-1}\left(\frac{\pi}{2}\right) - \frac{1}{2}\tan^{-1}\left(\frac{\pi}{4}\right) = 0.502 - 0.333 \approx 0.169$

304. **(C)** Since $f(x) = \frac{1}{\sqrt{16 - x^2}}$ has an infinite discontinuity at $x = 4,$ the integral is

improper. Evaluate $\int_{0}^{4} \frac{1}{\sqrt{16 - x^2}} dx = \lim_{k \to 4^-} \int_{0}^{k} \frac{1}{\sqrt{16 - x^2}} dx = \lim_{k \to 4^-} \int_{0}^{k} \frac{1}{\sqrt{4^2 - x^2}} dx =$

$\lim_{k \to 4^-} \sin^{-1}\left(\frac{x}{4}\right)\Big]_{0}^{k}.$ Since the limit exists, $\int_{0}^{4} \frac{1}{\sqrt{16 - x^2}} dx = \frac{\pi}{2}.$

305. **(D)** $\int_{0}^{2\pi} (x - \cos x)dx = \frac{x^2}{2} - \sin x \Big]_{0}^{2\pi} = \frac{4\pi^2}{2} - \sin(2\pi) - 0 + \sin 0 = 2\pi^2 - 0 - 0 + 0 = 2\pi^2$

306. **(D)** $\int_{0}^{k} (5 - x)dx = 5x - \frac{1}{2}x^2 \Big]_{0}^{k} = k^2 - 3k - 0.$ Set $-\frac{1}{2}k^2 + 5k = -12.$ Then $k^2 - 10k -$

$24 = 0.$ Factoring, we get $(k + 2)(k - 12) = 0 \Rightarrow k = -2, 12.$ Since the problem states that $k > 0,$ $k = 12.$

307. (A) $\displaystyle\sum_{i=1}^{n} n^2 i(i+2) = n^2 \sum_{i=1}^{n}(i^2 + 2i) = n^2 \sum_{i=1}^{n} i^2 + 2n^2 \sum_{i=1}^{n} i = n^2 \left[\frac{n(n+1)(2n+1)}{6}\right] +$

$2n^2\left[\dfrac{n(n+1)}{2}\right] = \dfrac{1}{6}n^3(n+1)(2n+1) + n^3(n+1) = n^3(n+1)\left[\dfrac{1}{3}n + \dfrac{7}{6}\right]$

308. (C) Let $u = 2x$, $du = 2dx \Rightarrow dx = \dfrac{du}{2}$. Substitute $\displaystyle\int \sin(2x)dx = \int \frac{1}{2}\sin u\, du =$

$-\dfrac{1}{2}\cos u + C$. Rewrite with x: $\displaystyle\int_{0}^{\pi/2} \sin(2x)dx = -\frac{1}{2}\cos 2x\,\Big]_{0}^{\pi/2} = -\frac{1}{2}(-1) + \frac{1}{2}(1) = \frac{1}{2} + \frac{1}{2} = 1.$

309. (B) $y = \displaystyle\int_{x}^{x^2}(t^2 - t + 1)dt = \frac{1}{3}t^3 - \frac{1}{2}t^2 + t\,\Big]_{x}^{x^2} = \frac{1}{3}x^6 - \frac{1}{2}x^4 + x^2 - \frac{1}{3}x^3 + \frac{1}{2}x^2 - x =$

$\dfrac{1}{3}x^6 - \dfrac{1}{2}x^4 - \dfrac{1}{3}x^3 + \dfrac{3}{2}x^2 - x. \quad \dfrac{dy}{dx} = 2x^5 - 2x^3 - x^2 + 3x - 1$

310. (B) Set $2x - 4 = 0$; $x = 2$; thus $|2x - 4| = \{2x - 4 \text{ if } x \geq 2, -(2x - 4) \text{ if } x < 2\}$. Rewrite $\int_{0}^{6}|2x - 4|\,dx = \int_{0}^{2}-(2x - 4)\,dx + \int_{2}^{6}(2x - 4)\,dx = [-x^2 + 4x]_{0}^{2} + [x^2 - 4x]_{2}^{6} = -(2)^2 + 4(2) - 0 + 6^2 - 4 + 8 = 20.$

311. (C) Let $u = \cos x + 1$, $du = -\sin x\,dx$. Rewrite: $-\displaystyle\int_{0}^{\pi}\frac{du}{u}$. Integrate: $-\ln|u|]_{0}^{\pi/2} =$
$-\ln|\cos x + 1|]_{0}^{\pi/2} = -\ln|0 + 1| + \ln|1 + 1| = \ln|2| - \ln|1| = \ln|2| \approx 0.693.$

312. (C) The limit $\displaystyle\lim_{k \to \infty}\int_{-k}^{0} e^x dx = 1 - e^{-k}$ which approaches 1 as $k \to \infty$, so that the integral is convergent and $\displaystyle\int_{-\infty}^{0} e^x dx = 1.$

313. (B) $\displaystyle\int_{-4}^{k}(2x - 3)dx = x^2 - 3x]_{-4}^{k} = k^2 - 3k - 16 + 3(-4)$. Set $k^2 - 3k - 28 = -30 \Rightarrow$
$k^2 - 3k + 2 = 0$. Factor the polynomial to find $(k - 2)(k - 1) = 0$. Therefore,
$$k = 1, 2$$
but $k > 1$ so $k = 2.$

314. (C) $y = \displaystyle\int_{x}^{2x}\frac{1}{\sqrt{t}}dt = \int_{x}^{2x}t^{-1/2}dt = 2t^{1/2}]_{x}^{x^2} = 2(x^2)^{1/2} - 2(x)^{1/2} = 2x - 2\sqrt{x} = 2(x - \sqrt{x}).$

Then $\dfrac{dy}{dx} = \dfrac{d}{dx}(2x - 2x^{1/2}) = 2 - x^{-1/2} = 2 - \dfrac{1}{\sqrt{x}}.$

315. (E) $\displaystyle\int_{-2}^{2}(x^2 - 2x + 1)dx = \frac{1}{3}x^3 - x^2 + x\,\Big]_{-2}^{2} = \frac{1}{3}(8) - 4 + 2 - \frac{1}{3}(-8) + 4 + 2 = \frac{16}{3} +$

$\dfrac{12}{3} = 9\dfrac{1}{3}$

316. (D) $\sum_{i=0}^{6} i^2(i-1) = \sum_{i=0}^{6} i^3 - i^2 = \dfrac{n^2(n+1)^2}{4} - \dfrac{n(n+1)(2n+1)}{6} = \dfrac{36(49)}{4} -$

$\dfrac{6(7)(13)}{6} = 441 - 91 = 350$

317. (C) $\displaystyle\int_0^{\pi/2} \sin^2 x\,dx = \dfrac{1}{2}\int_0^{\pi/2}(1-\cos 2x)\,dx = \dfrac{1}{2}\int_0^{\pi/2} dx - \dfrac{1}{2}\int_0^{\pi/2}\cos 2x\,dx =$

$\dfrac{x}{2} - \dfrac{\sin 2x}{4}\Big]_0^{\pi/2} = \dfrac{\pi}{4}$

318. (E) Let $u=x+1$. Then $x=u-1$ and $dx=du$. Therefore $\displaystyle\int_0^1 x(x+1)^{1/3}\,dx = \int_0^1 (u-1)u^{1/3}\,du =$

$\displaystyle\int_0^1 (u^{4/3} - u^{1/3})\,du = \int_0^1 u^{4/3}\,du - \int_0^1 u^{1/3}\,du$. Rewrite in terms of x: $\dfrac{3}{7}(x+1)^{7/3}\Big]_0^1 -$

$\dfrac{3}{4}(x+1)^{4/3}\Big]_0^1 = \dfrac{3}{7}(2)^{7/3} - \dfrac{3}{4}(2)^{4/3} = \dfrac{3}{7}\sqrt[3]{128} - \dfrac{3}{4}\sqrt[3]{16} \approx 0.591.$

319. (D) Set $x^2 - 9 = 0$; $x = \pm 3$. Thus $|x^2 - 9| = \{x^2 - 9 \text{ if } x \geq 3 \text{ or } x \leq 3, -(x^2 - 9) \text{ if } -3 <$

$x < 3\}$. Thus, $\displaystyle\int_{-6}^{6}|x^2 - 9|\,dx = \int_{-6}^{-3}(x^2-9)\,dx + \int_{-3}^{3} -(x^2-9)\,dx + \int_{3}^{6}(x^2-9)\,dx =$

$\dfrac{1}{3}x^3 - 9x\Big]_{-6}^{-3} - \left(\dfrac{1}{3}x^3 - 9x\right)\Big]_{-3}^{3} + \left(\dfrac{1}{3}x^3 - 9x\right)\Big]_{3}^{6} = \dfrac{1}{3}(-27) + 27 - \dfrac{1}{3}(-216) - (-9)(-6) -$

$\dfrac{1}{3}(27) + 27 - \dfrac{1}{3}(27) + 27 + \dfrac{1}{3}(216) - 54 - \dfrac{1}{3}(27) + 27 = 4(27) - 2(54) - \dfrac{4}{3}(27) +$

$\dfrac{2}{3}(216) = 144 - 36 = 108.$

320. (A) Let $u = \sqrt{x}$, $x = u^2$, $du = \dfrac{1}{2\sqrt{x}}dx \Rightarrow dx = 2\sqrt{x}\,du$. Then $2\displaystyle\int_1^4 e^u\,du = 2e^u\,]_1^4.$

Rewrite in terms of x: $\displaystyle\int_1^4 \dfrac{e^{\sqrt{x}}}{\sqrt{x}}\,dx = 2e^{\sqrt{x}}\Big]_1^4 = 2e^2 - 2e^1 = 2e(e-1).$

321. (D) Let $u = x^2$, $du = 2x\,dx \Rightarrow \dfrac{du}{2} = x\,dx$. Therefore, $\dfrac{1}{2}\displaystyle\int e^u\,du = \dfrac{1}{2}e^u + C$. Substitute

x: $\displaystyle\int_{\ln 2}^{\ln 3} xe^{x^2}\,dx = \dfrac{1}{2}e^{x^2}\Big]_{\ln 2}^{\ln 3} = \dfrac{1}{2}e^{(\ln 3)^2} - \dfrac{1}{2}e^{(\ln 2)^2} = 1.672 - 0.808 = 0.864.$

322. (C) Set $4x - 2 = 0 \Rightarrow x = \dfrac{1}{2}$. Thus $|4x-2| = \{4x-2 \text{ if } x \geq 2, -(4x-2) \text{ if } x < 2\}.$

Rewrite integral: $\displaystyle\int_0^{1/2} -(4x-2)\,dx + \int_{1/2}^{3}(4x-2)\,dx = -\dfrac{4x^2}{2} + 2x\Big]_0^{1/2} + \dfrac{4x^2}{2} - 2x\Big]_{1/2}^{3} =$

$-\dfrac{4\left(\dfrac{1}{4}\right)}{2} + 2\left(\dfrac{1}{2}\right) - 0 + \dfrac{4(9)}{2} - 2(3) - \dfrac{4\left(\dfrac{1}{4}\right)}{2} + 2\left(\dfrac{1}{2}\right) = 12 + 2 - 1 = 13.$

323. **(C)** Let $u = \tan x$, $du = \sec^2 x$. Then $\int u^2 \, du = \dfrac{u^3}{3} + C$. Then you can re-substitute x in

the definite integral: $\displaystyle\int_0^{\pi/3} \tan^2 x \sec^2 x \, dx = \dfrac{1}{3} \tan^3 x \Big]_0^{\pi/3} = \dfrac{1}{3}\left(\tan^3\left(\dfrac{\pi}{3}\right) - \tan^3 0\right) = \dfrac{(\sqrt{3})^3}{3} =$

$\dfrac{3\sqrt{3}}{3} = \sqrt{3}$.

324. **(A)** $\displaystyle\int_0^k \left(2x - \dfrac{\sqrt{x}}{2}\right) dx = 2\left(\dfrac{1}{2}\right)x^2 - \left(\dfrac{1}{2}\right)\dfrac{2x^{3/2}}{3}\Big]_0^k = 0$. Evaluating, you get $k^2 - \dfrac{1}{3}k^{3/2} = 0$.

Solving for k, you get $k^{1/2} = \dfrac{1}{3}$ or $k = \dfrac{1}{9}$.

325. **(D)** Since $f(x)$ has an infinite discontinuity at $x = 2$, the integral is improper. Evaluate

$\displaystyle\int_2^3 \dfrac{1}{\sqrt{x-2}} \, dx = \lim_{k\to 2^+} \int_k^3 \dfrac{1}{\sqrt{x-2}} \, dx$. Let $u = x - 2$ and $du = dx$. Then $\int u^{-1/2}$ you can rewrite

the definite integral $\lim_{k\to 2} + \displaystyle\int_k^3 \dfrac{1}{\sqrt{x-2}} \, dx = \lim_{k\to 2} + 2\sqrt{x-2}\,\Big]_k^3 = 2\sqrt{3-2} = 2$.

326. **(B)** Let $u = 3x$, $du = 3dx \Rightarrow \dfrac{du}{3} = dx$. Rewrite: $\dfrac{1}{3}\displaystyle\int 10^u \, du = \dfrac{10^u}{3\ln 10} + C$. Substitute

x: $\displaystyle\int_0^1 10^{3x} \, dx = \dfrac{10^{3x}}{3\ln 10}\Big]_0^1 = \dfrac{10^3}{3\ln 10} - \dfrac{10^0}{3\ln 10} = 144.76 - 0.144 = 144.62$.

327. **(E)** Let $u = \ln x$, $du = \dfrac{1}{x} \, dx$. Then $\int \dfrac{1}{u} \, du = \ln |u| + C$. Now substitute x in the defi-

nite integral: $\displaystyle\int_e^{e^2} \dfrac{1}{x \ln x} \, dx = \ln |\ln |x|| \,\Big]_e^{e^2} = \ln |\ln e^2| - \ln |\ln e| = \ln 2 - 0 = \ln 2$.

328. **(A)** Let $u = x^2$, $du = 2x \, dx \Rightarrow \dfrac{du}{dx} = 2x$. Then $y = \dfrac{1}{2}\displaystyle\int_0^u \cos t \, dt$. Taking the derivative

of y, $\dfrac{dy}{dx} = \left(\dfrac{dy}{du}\right)\left(\dfrac{du}{dx}\right) = \dfrac{1}{2}\cos u(2x) = \dfrac{1}{2}\cos(x^2)(2x) = x\cos x^2$.

329. **(C)** $\displaystyle\int_4^9 \dfrac{x+1}{\sqrt{x}} \, dx = \int_4^9 (x^{1/2} + x^{-1/2}) \, dx = \dfrac{x^{3/2}}{\frac{3}{2}} + \dfrac{x^{1/2}}{\frac{1}{2}}\Big]_4^9 = \dfrac{2}{3}(9)^{3/2} - \dfrac{2}{3}(4)^{3/2} + 2(9)^{1/2} -$

$2(4)^{1/2} = \dfrac{2}{3}(27) - \dfrac{2}{3}(8) + 2(3) - 2(2) = 14\dfrac{2}{3}$

330. (D) $\int_{-\infty}^{\infty} \frac{1}{1+x^2} dx = \int_{-\infty}^{0} \frac{1}{1+x^2} dx + \int_{0}^{\infty} \frac{1}{1+x^2} dx$. Evaluate each part of the integral

separately. $\int_{0}^{\infty} \frac{1}{1+x^2} dx = \lim_{k \to \infty} \int_{0}^{k} \frac{1}{1+x^2} dx = \lim_{k \to \infty} \tan^{-1}(x)]_0^k = \lim_{k \to \infty} [\tan^{-1}(k) -$

$\tan^{-1}(0)] = \frac{\pi}{2}$. $\int_{-\infty}^{0} \frac{1}{1+x^2} dx = \lim_{l \to -\infty} \int_{l}^{0} \frac{1}{1+x^2} dx = \lim_{l \to -\infty} \tan^{-1}(x)]_l^0 =$

$\lim_{l \to -\infty} [\tan^{-1}(0) - \tan^{-1}(l)] = \frac{\pi}{2}$. Then $\int_{-\infty}^{\infty} \frac{1}{1+x^2} = \frac{\pi}{2} + \frac{\pi}{2} = \pi$.

331. (B) $f'(x) = \tan^{-1} t]_0^x = \tan^{-1}(x) - \tan^{-1}(0)$. $f'\left(\frac{\pi}{6}\right) = \tan^{-1}\left(\frac{\pi}{6}\right) - \tan^{-1}(0) = 0.4823$.

Alternatively, you can integrate to find $f(x) = t \tan^{-1} t - \frac{1}{2} \ln(t^2 + 1)\Big]_0^x = x \tan^{-1} x -$

$\frac{1}{2} \ln(x^2 + 1)$. Then $f'(x) = \tan^{-1} x + \frac{x}{1+x^2} - \left(\frac{1}{2}\right)\frac{2x}{1+x^2} = \tan^{-1} x$.

332. (C) $\int_{-1}^{1} (x^2 - 3)(x^5 + 2) dx = \int_{-1}^{1} (x^7 - 3x^5 + 2x^2 - 6) dx = \frac{x^8}{8} - \frac{3x^6}{6} + \frac{2x^3}{3} - 6x\Big]_{-1}^{1} =$

$\frac{1}{8} - \frac{1}{2} + \frac{2}{3} - 6 - \left[\frac{1}{8} - \frac{1}{2} - \frac{2}{3} + 6\right] = -10\frac{2}{3}$

333. (E) $\int_{0}^{\infty} \frac{1}{x^2 - 3x + 2} dx = \int_{0}^{\infty} \frac{1}{(x-2)(x-1)} dx = \lim_{k \to \infty} \int_{0}^{k} \frac{1}{(x-2)(x-1)} dx$. Solve

integral by using partial fractions. Set $\frac{1}{(x-2)(x-1)} = \frac{A}{x-2} + \frac{B}{x-1}$. Then $A(x-1) +$

$B(x-2) = 1$. Therefore $Ax + Bx = 0$ and $-A - 2B = 1$. Solving for A and B, you find that

$A = 1, B = -1$. Now you have $\lim_{k \to \infty} \int_{0}^{k} \frac{1}{(x-2)(x-1)} dx = \lim_{k \to \infty} \int_{0}^{k} \frac{1}{x-2} dx +$

$\lim_{k \to \infty} \int_{0}^{k} \frac{1}{x-1} dx = \lim_{k \to \infty} [\ln |x-2|]_0^k + \ln |x-1|]_0^k] = \lim_{k \to \infty} \ln[(k-2)(k-1)] -$

$\ln[(0-1)(0-2)] = \infty$.

334. (C) Let $u = x^3$, $du = 3x^2 dx \Rightarrow \frac{du}{3} = x^2 dx$. Therefore, $\frac{1}{3} \int \frac{1}{\sqrt{1-u^2}} du = \frac{1}{3} \sin^{-1} u + C$.

Re-substitute x: $\int_{0}^{1} \frac{x^2}{\sqrt{1-x^6}} dx = \frac{1}{3} \sin^{-1} x^3\Big]_0^1 = \frac{1}{3} \sin^{-1}(1) - \frac{1}{3} \sin^{-1}(0) = \frac{1}{3}\left(\frac{\pi}{2}\right) = \frac{\pi}{6}$.

335. (A) $\frac{dy}{dx} = \frac{d}{dx} \int_{\sin x}^{\cos x} \left(1 - \frac{1}{2} t\right) dt = \left(1 - \frac{1}{2} \cos x\right)\frac{d}{dx}(\cos x) - \left(1 - \frac{1}{2} \sin x\right)\frac{d}{dx}(\sin x) =$

$-\sin x\left(1 - \frac{1}{2} \cos x\right) - \cos x\left(1 - \frac{1}{2} \sin x\right) = -\sin x + \frac{1}{2} \sin x \cos x - \cos x + \frac{1}{2} \sin x \cos x =$

$\sin x \cos x - \sin x - \cos x$.

336. (D) Let $u = \cot x$, $du = -\csc^2 x\,dx$. Solve the indefinite integral: $-\int u\,du = -\dfrac{u^2}{2} + C$.

Replace u with x: $\displaystyle\int_{\pi/6}^{\pi/4} \csc^2 x \cot x\,dx = -\dfrac{\cot^2 x}{2}\Bigg]_{\pi/6}^{\pi/4} = -\dfrac{\cot^2\left(\dfrac{\pi}{4}\right)}{2} + \dfrac{\cot^2\left(\dfrac{\pi}{6}\right)}{2} = -\dfrac{1}{2} + \dfrac{3}{2} = 1$.

337. (E) Let $u = \ln x$, $du = \dfrac{1}{x}\,dx$. Rewrite and solve: $\dfrac{1}{5}\int u\,du = \dfrac{1}{5}\left(\dfrac{u^2}{2}\right) + C$. Substitute x

into the definite integral: $\displaystyle\int_1^e \dfrac{\ln x}{5x}\,dx = \dfrac{1}{5}\dfrac{(\ln x)^2}{2}\Bigg]_1^e = \dfrac{1}{10}(\ln e)^2 - \dfrac{1}{10}(\ln 1)^2 = \dfrac{1}{10}$.

338. (B) $\displaystyle\int_0^9\left(\dfrac{1}{\sqrt{x}} - k\right)dx = \int_0^9 (x^{-1/2} - k)\,dx = \dfrac{x^{1/2}}{\dfrac{1}{2}} - kx\Bigg]_0^9 = 2\sqrt{x} - kx\Bigg]_0^9 = 12$. Evaluating,

you get $6 - 9k = 12$ or $k = -\dfrac{2}{3}$.

339. (D) Let $u = 2x$, $du = 2dx \Rightarrow \dfrac{du}{2} = dx$. Rewrite: $\dfrac{1}{2}\int g(u)\,du = \dfrac{1}{2}f(u)$. Then

$\displaystyle\int_0^{2\pi} g(2x)\,dx = \dfrac{1}{2}f(2x)\Bigg]_0^{\pi} = \dfrac{1}{2}[f(2\pi) - f(0)]$.

340. (A) Let $u = e^x + 4$, $du = e^x dx$. Rewrite: $\displaystyle\int \dfrac{1}{u}\,du = \ln|u| + C$. Re-substitute x:

$\displaystyle\int_{\ln 3}^{\ln 5} \dfrac{e^x}{e^x + 4}\,dx = \ln|e^x + 4|\Bigg]_{\ln 3}^{\ln 5} = \ln|e^{\ln 5} + 4| - \ln|e^{\ln 3} + 4| = \ln(9) - \ln(7) = \ln\left(\dfrac{9}{7}\right)$.

341. (C) Let $u = \sin x$, $du = \cos x\,dx$. Rewrite: $\displaystyle\int u^{1/2}\,du = \dfrac{2u^{3/2}}{3} + C$. Now substitute x:

$\displaystyle\int_0^{\pi/6} \sqrt{\sin x}\cos x\,dx = \dfrac{2(\sin x)}{3}\Bigg]_0^{\pi/6} = \dfrac{2\left(\sin\dfrac{\pi}{6}\right)^{3/2}}{3} - 0 = \dfrac{2\left(\dfrac{1}{2}\right)^{3/2}}{3} = \dfrac{2\sqrt{\dfrac{1}{8}}}{3} = \dfrac{\sqrt{8}}{12} = \dfrac{\sqrt{2}}{6}$.

342. (C) $G(x) = \displaystyle\int \ln(x)\,dx = x\ln|x| - x + C = 1\ln 1 - 1 + C = 0 \Rightarrow C = 1$. $G(2) = 2\ln 2 - 2 + 1 = 2\ln 2 - 1$.

343. (B) Use integration by parts letting $u = e^x$ and $dv = \sin x\,dx$. Then $du = e^x dx$, $v = -\cos x$. Then $\int e^x \sin x\,dx = -e^x \cos x + \int e^x \cos x\,dx$. Use integration by parts a second time. Let $u = e^x$, $dv = \cos x\,dx$. Then $du = e^x dx$, $v = \sin x$. Now $\int e^x \sin x\,dx = -e^x \cos x + e^x \sin x - e^x \sin x$. Combine terms to get $2\int e^x \sin x\,dx = -e^x \cos x + e^x \sin x$. Then, dividing both sides by 2, we get $\int e^x \sin x\,dx = \dfrac{1}{2}e^x \sin x - \dfrac{1}{2}e^x \cos x$. Now evaluate the integral including the limits of integration. You now have $\int_0^{\pi/2} e^x \sin x\, dx = \frac{1}{2} e^x \sin x \Big]_0^{\pi/2} - \frac{1}{2} e^x \cos x \Big]_0^{\pi/2} =$

$\frac{1}{2} e^{\pi/2} \left(\sin \frac{\pi}{2} - \cos \frac{\pi}{2} \right) - \frac{1}{2} e^0 (\sin 0 - \cos 0) = \frac{1}{2} e^{\pi/2} - \frac{1}{2}(-1) = \frac{1}{2}(e^{\pi/2} + 1).$

344. (B) $\int_6^{10} \frac{1}{x^2 - 3x - 10}\, dx = \int_6^{10} \frac{1}{(x-5)(x+2)}\, dx.$ Use integration by partial fractions. Set

$\frac{1}{(x-5)(x+2)} = \frac{A}{x-5} + \frac{B}{x+2}.$ Then $A(x+2) + B(x-5) = 1.$ So, $Ax + Bx = 0,\ 2A - 5B = 1.$

Solving for A and B, you get $A = \frac{1}{7},\ B = -\frac{1}{7}.$ You now have $\int_6^{10} \frac{dx}{x^2 - 3x - 10} = \int_6^{10} \frac{\frac{1}{7}}{x-5}\, dx -$

$\int_6^{10} \frac{\frac{1}{7}}{x+2}\, dx = \frac{1}{7} \ln |x - 5| - \frac{1}{7} \ln |x + 2| \Big]_6^{10} = \frac{1}{7}[\ln 5 - \ln 12 - \ln 1 + \ln 8] = \frac{1}{7} \ln \left| \frac{8 \cdot 5}{12 \cdot 1} \right| =$

$\frac{1}{7} \ln \left| \frac{10}{3} \right|.$

345. (C) Evaluate in the limit, $\lim_{k \to \infty} \int_1^k \frac{1}{x^6}\, dx = \lim_{k \to \infty} -\frac{x^{-5}}{5} \Big]_1^k = \lim_{k \to \infty} \frac{(-x^{-5})}{5} \Big]_1^k.$

As $k \to \infty,\ \frac{1}{k^5} \to 0$ and $\lim_{k \to \infty} \left[\frac{1}{5} - \frac{1}{5k^5} + \frac{1}{5} - \frac{1}{5} \right]$ converges to $\frac{1}{5}.$

346. (A) $C(x) = \int C'(x) dx = \int \left(\frac{1}{4} x - 2 \right) dx = \frac{1}{8} x^2 - 2x + C = \frac{1}{8} x^2 - 2x + 2$

(B) $\bar{C} = \frac{C(x)}{x} = \frac{1}{8} x - 2 + \frac{2}{x}.$ Set $\frac{d\bar{C}}{dx} = \frac{1}{4} x - \frac{1}{x^2} = 0.$ Solving for x, you find $x^3 = 4$

or $x = \sqrt[3]{4}.$ Find the second derivative at this point to test whether this is a relative

minimum or maximum. $\frac{d^2 \bar{C}}{dx^2} = \frac{1}{4} + \frac{1}{3x^3}.$ At $x = \sqrt[3]{4},\ \frac{d^2 \bar{C}}{dx^2} \Big|_{x=\sqrt[3]{4}} = \frac{1}{4} + \frac{1}{12} = \frac{1}{3} > 0$ so

$C(\sqrt[3]{4})$ is a relative minimum.

(C) $C(40) = \int_0^{40} C'(x) dx = \int_0^{40} \left(\frac{1}{4} x - 2 \right) dx = \frac{x^2}{8} - 2x \Big]_0^{40} = \frac{1}{8}(40)^2 - 2(40) - 0 =$

$600 - 80 = 520.$

347. (A) The size of the subdivision is given by $\Delta x_i = \frac{4-0}{4} = 1.$ Then make a table of values for $f(x).$

x	0	1/2	1	3/2	2	5/2	3	7/2	4
$f(x)$	0	1/4	1	9/4	4	25/4	9	49/4	16

Now compute $A = \sum_{i=1}^{4} f(c_i) \Delta x_i = 1(1) + 4(1) + 9(1) + 16(1) = 30$

(B) Now $\Delta x_i = \dfrac{4-0}{8} = \dfrac{1}{2}$. Then $A = \sum_{i=1}^{8} f(c_i)\Delta x_i = \dfrac{1}{4}\left(\dfrac{1}{2}\right)+1\left(\dfrac{1}{2}\right)+\dfrac{9}{4}\left(\dfrac{1}{2}\right)+$

$4\left(\dfrac{1}{2}\right)+\dfrac{25}{4}\left(\dfrac{1}{2}\right)+9\left(\dfrac{1}{2}\right)+\dfrac{49}{4}\left(\dfrac{1}{2}\right)+16\left(\dfrac{1}{2}\right)=\dfrac{1}{8}+\dfrac{4}{8}+\dfrac{9}{8}+\dfrac{16}{8}+\dfrac{25}{8}+\dfrac{36}{8}+$

$\dfrac{49}{8}+\dfrac{64}{8}=25\dfrac{1}{2}$.

(C) The area $A = \displaystyle\int_0^4 x^2\,dx = \dfrac{1}{3}x^3\Big]_0^4 = \dfrac{64}{3}-0=21\dfrac{1}{3}$.

348. (A) $P(t) = \displaystyle\int (10t - 2t^{1/2} + 100)dt = \dfrac{10t^2}{2} - \dfrac{2t^{3/2}}{\frac{3}{2}} + 100t + C$. Evaluate at $P(0)$ using

your initial condition to find C. $P(0) = 5(0)+\dfrac{4}{3}(0)+100(0)+C=500 \Rightarrow C=500$. Then you

can replace C in our equation to find $P(t) = 5t^2 - \dfrac{4}{3}t^{3/2} + 100t + 500$.

(B) $P(t) = \displaystyle\int_0^3 (10t - 2t^{1/2}+100)dt = 5t^2 - \dfrac{4}{3}t^{3/2} + 100t + 500\Big]_0^3 = 5(9) - \dfrac{4}{3}(3\sqrt{3}) +$

$300 + 500 - 500 = 338$.

(C) $P_{avg.} = \dfrac{P(12) - P(0)}{12} = \dfrac{1}{12}\left[5(144) - \dfrac{4(\sqrt{1728})}{3} + 1200 + 500 - 500\right]$

$\approx \dfrac{1}{12}(720 - 111 + 1200) \approx 151$ per hour.

349. (A) $v(t) = \displaystyle\int a(t)dt = \int 15t^{1/2}dt = 15\dfrac{t^{3/2}}{\frac{3}{2}} + C = 15\left(\dfrac{2}{3}\right)t^{3/2} + C = 10\sqrt{t^3} + C$. Use

the initial condition for $v(0)$ to find that $v(0) = 10(0) + C = 50 \Rightarrow C = 50$. Therefore
$v(t) = 10\sqrt{t^3} + 50$.

(B) $s(t) = \displaystyle\int v(t)dt = \int (10\sqrt{t^3} + 50)dt = \int 10t^{3/2}dt + \int 50dt = 10\left(\dfrac{2}{5}\right)t^{5/2} + 50t + C =$

$4t^{5/2} + 50t + C$. Use the initial condition for position to find a value for C. You
know $s(0) = 4(0) + 50(0) + C = 100$. Solving for C, you get $C = 100$. Therefore
$s(t) = 4t^{5/2} + 50t + 100$.

(C) Use the result from part A over the interval from 0 to 5 to write $v(t) = \displaystyle\int_0^5 a(t)dt =$

$10\sqrt{t^3} + 50\Big]_0^5 = (10\sqrt{125} + 50 - 0 - 50) = 50\sqrt{5} \approx 111$.

(D) Use the result from part B over the interval from 0 to 5 to write $s(t) = \displaystyle\int_2^5 v(t)dt =$

$4t^{5/2} + 50t + 100\Big]_0^5 = 4(\sqrt{3125}) + 250 + 100 - 0 - 100 \approx 474$.

350. (A) $m(x) = \int_0^{35} \rho(x)dx = \frac{1}{3}\int_0^{35} \frac{1}{(x+1)^{1/2}} dx$. Let $u = x+1$, $du = dx$. Rewrite:

$$\frac{1}{3}\int_0^{35} u^{-1/2} du = \frac{1}{3}(2)(u^{1/2})\Big]_0^{35} = \frac{2}{3}(x+1)^{1/2}\Big]_0^{35} = 4 - \frac{2}{3} = \frac{12}{3} - \frac{2}{3} = 3\frac{1}{3} \text{ tons.}$$

(B) $\int_0^b \rho(x)dx = \frac{1}{2}\left(3\frac{1}{3}\right)$ tons. $\frac{2}{3}(x+1)^{1/2}\Big]_0^b = \frac{10}{6}$ tons. $\frac{2}{3}(b+1)^{1/2} - \frac{2}{3} = \frac{10}{6}$ tons.

$(b+1)^{1/2} = \left(\frac{10}{6} + \frac{2}{3}\right)\frac{3}{2} \Rightarrow (b+1)^{1/2} = \frac{7}{2} \Rightarrow b+1 = \frac{49}{4} \Rightarrow b = \frac{45}{4} = 11\frac{1}{4}$ feet.

(C) $m = \int_{30}^{35} \rho(x)dx = \frac{2}{3}(x+1)^{1/2}\Big]_{30}^{35} = \frac{2}{3}(36)^{1/2} - \frac{2}{3}(31)^{1/2} \approx 0.29$ tons.

Chapter 8: Areas and Volumes

351. (B) The exact area is computed

$$\int_0^3 x^2 + 2dx = \left[\frac{x^3}{3} + 2x\right]_0^3 = 9 + 6 = 15$$

The approximation using 3 midpoint rectangles is computed

$$\sum_{i=1}^3 f(x_i)\left(\frac{3-0}{3}\right) = f\left(\frac{1}{2}\right)\cdot 1 + f\left(\frac{3}{2}\right)\cdot 1 + f\left(\frac{5}{2}\right)\cdot 1 = \frac{9}{4} + \frac{17}{4} + \frac{33}{4} = \frac{59}{4} \text{ or } 14.75$$

where x_i is the midpoint of the ith subinterval of $[0, 3]$. The difference between the approximation and the actual area is therefore $15 - 14.75 = 0.25$ or $\frac{1}{4}$.

352. (E) The area of this region may be attained by computing

$$\int_0^{\pi/2} |\sin x - \cos x| \, dx$$

Note that

$$|\sin x - \cos x| = \begin{cases} \cos x - \sin x, & \text{if} \quad 0 \le x \le \frac{\pi}{4} \\ \sin x - \cos x, & \text{if} \quad \frac{\pi}{4} < x \le \frac{\pi}{2} \end{cases}$$

Thus,

$$\int_0^{\pi/2} |\sin x - \cos x| \, dx = \int_0^{\pi/4} (\cos x - \sin x)dx + \int_{\pi/4}^{\pi/2} (\sin x - \cos x)dx$$

$$= [\sin x + \cos x]_0^{\pi/4} + [-\cos x - \sin x]_{\pi/4}^{\pi/2}$$

$$= [\sqrt{2} - 1] + [-1 + \sqrt{2}] = 2\sqrt{2} - 2 \qquad \text{or} \qquad 2(\sqrt{2} - 1)$$

353. (D) Recall that $\tan x = \dfrac{\sin x}{\cos x}$. Then, when $u = \cos x$, you see that $\tan x \, dx = \dfrac{-du}{u}$.

Further, $u = 1$ when $x = 0$, and $u = 1/2$ when $x = \pi/3$. Therefore,

$$\int_{0}^{\pi/3} \tan x \, dx = \int_{1/2}^{1} \frac{+du}{u} = \Big[\ln |u|\Big]_{1/2}^{1} = \ln 1 - \ln\left(\frac{1}{2}\right) = \ln 2$$

354. (A) You must solve for b in the integral equation $2 = \displaystyle\int_{1}^{b} \frac{\ln x}{x} \, dx$. Under the substitution $u = \ln x$, the integral becomes

$$\int_{0}^{\ln b} u \, du = \left[\frac{u^2}{2}\right]_{0}^{\ln b} = \frac{(\ln b)^2}{2}$$

You have shown $2 = \dfrac{(\ln b)^2}{2}$. This reduces to $(\ln b)^2 = 4$, which is true only when $\ln b = 2$ or $\ln b = -2$. Exponentiating both equations base e, you obtain $b = e^2$ and $b = e^{-2}$. Since b is stated to be greater than 1, you see that it must be that $b = e^2$.

355. (C) The area of this region may be attained by computing

$$\int_{0}^{2} |f(x) - g(x)| \, dx$$

Using the graph, you see that

$$|f(x) - g(x)| = \begin{cases} f(x) - g(x), & \text{if} \quad 0 \le x \le 1 \\ g(x) - f(x), & \text{if} \quad 1 < x \le 2 \end{cases}$$

Thus,

$$\int_{0}^{2} |f(x) - g(x)| \, dx = \int_{0}^{1} (x^3 - 3x^2 + 2x) dx + \int_{1}^{2} 3x^2 - 2x - x^3 dx$$

$$= \left[\frac{x^4}{4} - x^3 + x^2\right]_{0}^{1} + \left[x^3 - x^2 - \frac{x^4}{4}\right]_{1}^{2} = \frac{1}{4} + \frac{1}{4} = \frac{1}{2}$$

356. (D) The region in question is shown in the figure on next page.

The area of this region may be attained by computing

$$\int_{0}^{1} \left(e^{-x} + 2 + \frac{x}{2}\right) dx = \left[-e^{-x} + 2x + \frac{x^2}{4}\right]_{0}^{1} = \left(-\frac{1}{e} + 2 + \frac{1}{4}\right) + 1 = \frac{13}{4} - \frac{1}{e} = \frac{13e - 4}{4e}$$

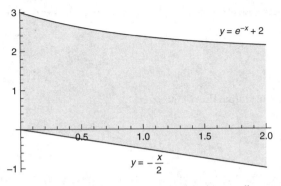

The region bounded by the graphs of $y = e^{-x} + 2$ and $y = -\dfrac{x}{2}$.

357. (E) It is clear from the unit circle that sine is at least than $1/2$ on the interval $[\pi/6, 5\pi/6]$. Therefore, the solutions $x \in [0, 2\pi]$ to the equation $2\sin x \geq 1$ belong only to $[\pi/6, 5\pi/6]$, and you attain the area in question by computing

$$\int_{\pi/6}^{5\pi/6} 2\sin x - 1\,dx = [-2\cos x - x]_{\pi/6}^{5\pi/6} = \left[\sqrt{3} - 5\pi/6\right] - \left[-\sqrt{3} - \pi/6\right] = 2\left(\sqrt{3} - \frac{\pi}{3}\right)$$

358. (B) You need to evaluate the definite integral $\int_1^e 2x\ln x\,dx$. You proceed by integration-by-parts, setting $u = \ln x$ and $dv = 2x\,dx$ in the formula $\int u\,dv = uv - \int v\,du$ to obtain

$$\int_1^e 2x\ln x\,dx = [x^2\ln x]_1^e - \int_1^e x\,dx = e^2 - \left[\frac{x^2}{2}\right]_1^e = e^2 - \left(\frac{e^2}{2} - \frac{1}{2}\right) = \frac{1}{2}(e^2 + 1)$$

359. (D) From the half-angle identity for cosine

$$\cos^2(x) = \frac{1 + \cos(2x)}{2}$$

and the observation that $y = \cos^2(x)$ is symmetric about the y-axis, you attain the area of the region in question by computing

$$2\int_0^{\pi/6} \cos^2(x)\,dx = \int_0^{\pi/6} 1 + \cos(2x)\,dx = \left[x + \frac{\sin(2x)}{2}\right]_0^{\pi/6} = \frac{\pi}{6} + \frac{\sqrt{3}}{4}$$

360. (D) You see that when $x = 4$, $f(x) = 2e^2 > 3$. Solving for x in the equation $2e^{x/2} = 3$, you see that $x = 2\ln\left(\dfrac{3}{2}\right) < 4$. Therefore, to find the area of the bounded region, you compute

$$\int_{2\ln\left(\frac{3}{2}\right)}^{4} 2e^{x/2} - 3\,dx = \left[4e^{x/2} - 3x\right]_{2\ln\left(\frac{3}{2}\right)}^{4} = [4e^2 - 12] - \left[6 - 6\ln\left(\frac{3}{2}\right)\right]$$

$$= 4e^2 + 6\ln\left(\frac{3}{2}\right) - 18$$

361. (C) From basic trigonometric identities and algebra you see that the expression simplifies to

$$\frac{\sec x + \csc x}{1 + \tan x} = \csc x$$

Therefore, the area is the value obtained from the computation:

$$\int_{\pi/6}^{\pi/3} \csc x \, dx = \left[\ln |\csc x - \cot x| \right]_{\pi/6}^{\pi/3}$$

$$= \ln \left(\frac{2}{\sqrt{3}} - \frac{1}{\sqrt{3}} \right) - \ln(2 - \sqrt{3}) = -\ln(\sqrt{3}) - \ln(2 - \sqrt{3}) \approx 0.768$$

362. (A) The area is given by the value of the integral

$$\int_{r}^{s} |f(x) - g(x)| \, dx$$

where $x = r$ and $x = s$ are the bounds of integration. To find these bounds, you solve for x in the equation $\ln x = 3 - 2 \ln x$. Adding $2 \ln x$ to both sides of the equation and dividing by 3 reduces the equation to $\ln x = 1$. Exponentiating this equation base e yields $x = e$. Therefore, your bounds are $r = e$ and $r = e^2$. Further, since $\ln x - 1 \geq 0$ when $x \geq e$, you see that $f(x) \geq g(x)$ for $x \in [e, e^2]$. Therefore, you compute the following integral:

$$3 \int_{e}^{e^2} \ln x - 1 \, dx = 3[x \ln x - 2x]_{e}^{e^2} = 3[(2e^2 - 2e^2) - (e - 2e)] = 3[0 - e] = 3e$$

363. (A) Upon consideration of the graphs of the functions in question, you see that the area is attained by computing

$$\int_{0}^{\theta} \cos(x) - \cos(2x) \, dx = \left[\sin(x) - \frac{1}{2} \sin(2x) \right]_{0}^{\theta} = \sin(\theta) - \frac{1}{2} \sin(2\theta)$$

where $0 < \theta \leq \pi$ is the nonzero solution to $\cos(x) = \cos(2x)$. By the double angle identity for cosine this is reduced to the equation $\cos(x) = 2 \cos^2(x) - 1$, which is quadratic in $\cos x$. Subtracting $\cos x$ from both sides yields $0 = 2 \cos^2(x) - \cos(x) - 1 = (2 \cos x + 1)(\cos x - 1)$. The only solution to this equation in $(0, \pi]$ is $\theta = 2\pi/3$, coming from solving $\cos x = -\frac{1}{2}$.

Therefore, the area of the bounded region is $\sin(2\pi/3) - \frac{1}{2}\sin(4\pi/3) = \frac{\sqrt{3}}{2} + \frac{\sqrt{3}}{4} = \frac{3\sqrt{3}}{4}$.

364. (B) Observe that

$$3A = 3 \int_{0}^{a} (x^2 + 1) \, dx = 3 \left[\frac{x^3}{3} + x \right]_{0}^{a} = a^3 + 3a \quad \text{and} \quad B = \frac{8a^3}{3} + 2a$$

Therefore, you solve $3A = B$ for a.

$$a^3 + 3a = \frac{8a^3}{3} + 2a \Leftrightarrow 3a^3 + 3a = 8a^3 \Leftrightarrow 0 = a(5a^2 - 3)$$

Therefore, since $a > 0$, the only solution is $a = \sqrt{\dfrac{3}{5}}$.

365. (B) Observe that

$$2A = 2\int_0^a \frac{1}{x}\, dx = 2[\ln |x|]_1^a = 2 \ln a \qquad \text{and} \qquad B = \ln b$$

Therefore, you solve $2A = B$ for b:

$$2 \ln a = \ln b \Leftrightarrow a^2 = b$$

366. (D) You solve for y in the equation $x^2 + \dfrac{y^2}{4} = 1$, to obtain $y = \pm 2\sqrt{1 - x^2}$. Since you are considering the region above $y = 1$, the area is obtained by integrating

$$\int_r^s 2\sqrt{1 - x^2} - 1\, dx = 2\int_r^s \sqrt{1 - x^2}\ dx - (s - r) = 2I - (s - r)$$

where $r = -\dfrac{\sqrt{3}}{2}$ and $s = \dfrac{\sqrt{3}}{2}$ are the solutions to $x^2 + \dfrac{y^2}{4} = 1$ when $y = 1$.

In order to compute the integral I appearing in the right-hand side above, you make the trigonometric substitution $x = \sin\theta$ in order to make use of the fact that $1 - \sin^2\theta = \cos^2\theta$. Then $dx = \cos\theta\, d\theta$, and the bounds become $\theta = m\pi/3$. Therefore,

$$I = \int_{-\frac{\sqrt{3}}{2}}^{\frac{\sqrt{3}}{2}} \sqrt{1 - x^2}\, dx = 2\int_0^{\pi/3} \cos^2\theta\, d\theta = 2\int_0^{\pi/3} \left(\frac{1 + \cos(2\theta)}{2}\right) d\theta$$

$$= \left[\theta + \frac{\sin(2\theta)}{2}\right]_0^{\pi/3} = \frac{\pi}{3} + \frac{\sqrt{3}}{4}$$

where you have also made use of the half-angle identity for cosine.

The area in question is therefore

$$2I - \left(\frac{\sqrt{3}}{2} - \left(-\frac{\sqrt{3}}{2}\right)\right) = \frac{2\pi}{3} + \frac{\sqrt{3}}{2} - \sqrt{3} = \frac{2\pi}{3} - \frac{\sqrt{3}}{2}$$

367. (D) The area is obtained upon computing

$$\int_0^3 h(x)\, dx = \int_0^3 xf(x^2)\, dx$$

You make the substitution $u = x^2$, so that $\frac{1}{2} du = x \, dx$ and integrate

$$\frac{1}{2}\int_0^9 f(u) \, du = \frac{1}{2}(g(9) - g(0))$$

368. (E) The area of the region is obtained by computing

$$\int_0^r \cos x - (e^x - 1) \, dx = [\sin x - e^x + x]_0^r = \sin r - e^r + r + 1$$

where $r > 0$ is the unique positive solution to $\cos x = e^x - 1$. Using a calculator to find the x-coordinate of this point of intersection yields $r \approx 0.6013467$. Therefore, substituting in for r, the area of the region is approximately 0.342526 which may be rounded to 0.343.

369. (C) Observe that $x = 1$ when $t = 0$, and $x = 2 + \ln 2$ when $t = \ln 2$. Therefore, the area under the curve is given by the integral

$$\int_1^{2+\ln 2} y \, dx = \int_0^{\ln 2} (1 + e^t)^2 \, dt = \int_0^{\ln 2} (1 + 2e^t + e^{2t}) \, dt$$

$$= \left[t + 2e^t + \frac{e^{2t}}{2} \right]_0^{\ln 2} = (\ln(2) + 4 + 2) - \left(2 + \frac{1}{2} \right) = \ln 2 - \frac{7}{2}$$

where you have used the substitution technique of integration, observing that $dx = 1 + e^t \, dt$

370. (E) The curve is shown in the figure below.

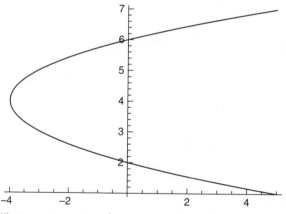

The parametric curve: $x = t^2 + 4t$ and $y = 2 - t$.

To find where the curve intersects the y-axis, you first solve $x(t) = 0$ to obtain $t = 0$ and $t = -4$. Noting that $dx = 2t + 4 \, dt$, the area is obtained by integrating

$$\int_{-4}^0 (2 - t)(2t + 4) \, dt = -2\int_{-4}^0 t^2 - 4 \, dt = -2\left[\frac{t^3}{3} - 4t \right]_{-4}^0 = -2\left[\frac{-64}{3} + 16 \right] = -2\left(\frac{-16}{3} \right) = \frac{32}{3}$$

This may also be computed by observing that $t = 0$ and $t = -4$ correspond to $y = 2$ and $y = 6$, respectively, and integrating

$$-\int_2^6 x\, dy = -\int_0^{-4} -(t^2 + 4t)\, dt = \left[\frac{t^3}{3} + 2t^2\right]_0^{-4} = \left[\frac{-64}{3} + 32\right] = \frac{32}{3}$$

where the negative sign appears in front of the initial integral since the curve being *to the left of* the y-axis when integrating with respect to y is analogous to the curve being *below* the x-axis when integrating with respect to x.

371. (E) The region in question is shown in the figure below.

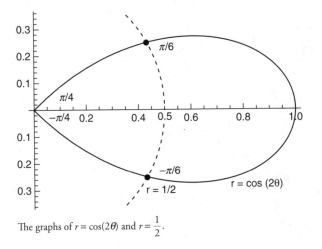

The graphs of $r = \cos(2\theta)$ and $r = \dfrac{1}{2}$.

You first observe that the only solutions in $[-\pi/4,\ \pi/4]$ to the equation $\cos 2\theta = 1/2$ is $\theta = \pm\pi/6$. We therefore see that the area may be found by subtracting the area inside the sector of the circle defined by $r = 1/2$ from the area inside the curve defined by $r = \cos 2\theta$ when we allow θ to vary from $-\pi/6$ to $\pi/6$. The area is given by the following integral:

$$\frac{1}{2}\int_{-\pi/6}^{\pi/6}\left(\cos^2(2\theta) - \left(\frac{1}{2}\right)^2\right)d\theta = \int_0^{\pi/6}\left(\frac{1 + \cos(4\theta)}{2} - \frac{1}{4}\right)d\theta$$

$$= \left[\frac{\theta}{2} + \frac{\sin(4\theta)}{8} - \frac{\theta}{4}\right]_0^{\pi/6} = \frac{\pi}{24} + \frac{\sqrt{3}}{16}$$

372. (B) The polar curve is shown in the figure below.

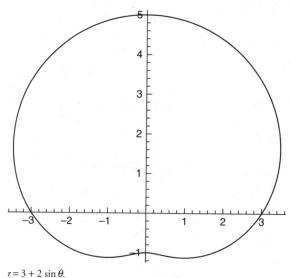

$r = 3 + 2 \sin \theta.$

The area of the right half of this curve is the result of integrating

$$\frac{1}{2} \int_{-\pi/2}^{\pi/2} r^2 d\theta = \frac{1}{2} \int_{-\pi/2}^{\pi/2} (3 + 2 \sin \theta)^2 \, d\theta$$

$$= \frac{1}{2} \int_{-\pi/2}^{\pi/2} (9 + 12 \sin \theta + 4 \sin^2 \theta) \, d\theta$$

$$= \frac{1}{2} \int_{-\pi/2}^{\pi/2} (9 + 12 \sin \theta + 2\{1 - \cos(2\theta)\}) \, d\theta$$

$$= \frac{1}{2} \left[9\theta - 12 \cos \theta + 2 \left\{ \theta - \frac{\sin(2\theta)}{2} \right\} \right]_{-\pi/2}^{\pi/2}$$

$$= \frac{1}{2}[(9\pi/2 + 2\{\pi/2\}) - (-9\pi/2 - 2\{\pi/2\})] = \frac{11\pi}{2}$$

373. (C) The length of each square cross-section is a function of x given by $s(x) = 2\sqrt{x}$, and therefore each square has area $A(x) = 4x$. The volume of the described solid is thus

$$\int_0^3 A(x) \, dx = [2x^2]_0^3 = 18$$

374. (A) First, observe that the points of intersection are the solutions to the equation $\frac{3}{x} = 4 - x$, which may be rearranged as $x^2 - 4x + 3 = 0$, and thus the solutions are $x = 1$ and $x = 3$ (these appear below as the bounds of integration). The base of each rectangular cross-section is a function of x given by $b(x) = 4 - x - \frac{3}{x}$, and therefore each rectangle has area $A(x) = 2\left(4 - x - \frac{3}{x}\right)$. Therefore, the volume of S is the value of the following integral:

$$\int_1^3 A(x)\,dx = 2\int_1^3 \left(4 - x - \frac{3}{x}\right)^2 dx \approx 0.6$$

where you have used a calculator to approximate the integral.

375. (C) The side of each triangle is a function of y given by $s(y) = 2 - 2^y$. It follows that the height of each triangle is given by $h(y) = \frac{\sqrt{3}}{2}(2 - 2^y)$, and its area, therefore, has the formula

$$A(y) = \frac{\sqrt{3}}{4}(2 - 2^y)^2$$

You compute the following integral to find the volume of the solid in question:

$$\int_{-1}^1 A(y)\,dy = \frac{\sqrt{3}}{4}\int_{-1}^1 (2 - 2^y)^2\,dy \approx 0.887$$

where the bound $y = 1$ is the value of $y = \log_2(x)$ when $x = 2$.

376. (E) Each cross-section perpendicular to the x-axis will be a circle of radius $r = \ln x$. Therefore, the volume of the solid is the value of the following integral:

$$\int_1^e \pi(\ln x)^2\,dx$$

You proceed to integrate by parts, setting $u = (\ln x)^2$ and $dv = dx$ in the formula $\int u\,dv = uv - \int v\,du$. Therefore,

$$\int_1^e \pi(\ln x)^2\,dx = \pi\left([x(\ln x)^2]_1^e - 2\int_1^e x\left(\frac{\ln x}{x}\right)dx\right)$$

$$= \pi\left([x(\ln x)^2]_1^e - 2[x\ln x - x]_1^e\right)$$

$$= \pi(e(\ln e)^2 - 2[(e\ln e - e) + 1]) = \pi(e - 2)$$

377. (A) Observe that the left- and right-halves of the unit circle are defined by the equations $\ell(y) = -\sqrt{1-y^2}$ and $r(y) = +\sqrt{1-y^2}$. Furthermore, observe that each cross-section of the described solid is that of a large disk with a smaller disk removed (called a "washer"), with area $A(y) = \pi(2-\ell)^2 - \pi(2-r)^2$ symmetry about the x- and y-axes, we see that $A(y) = 4[\pi(2-\ell)^2 - 4\pi] = 4\pi(1-y^2 + 4\sqrt{1-y^2})$ as y varies from 0 to 1. Therefore, the volume V of the solid is given by the integral:

$$V = 4\pi \int_0^1 (1-y^2 + 4\sqrt{1-y^2})\,dy = 4\pi\left(\left[y - \frac{y^3}{3}\right]_0^1 + 4\int_0^1 \sqrt{1-y^2}\,dy\right)$$

Let $I = \int_0^1 \sqrt{1-y^2}\,dy$. In order to evaluate I, we make the trigonometric substitution $y = \sin\theta$. Observe that under this substitution you have that $\sqrt{1-y^2} = \cos\theta$ and $dy = \cos\theta\,d\theta$; moreover, the θ-bounds of integration become $\theta = 0$ and $\theta = \pi/2$. Thus,

$$I = \int_0^{\pi/2} \cos^2\theta\,d\theta = \frac{1}{2}\int_0^{\pi/2} (1+\cos(2\theta))\,d\theta = \frac{1}{2}\left[\theta + \frac{\sin(2\theta)}{2}\right]_0^{\pi/2} = \frac{\pi}{4}$$

Therefore,

$$V = 4\pi\left(\left[y - \frac{y^3}{3}\right]_0^1 + 4I\right) = 4\pi\left(\frac{2}{3} + \pi\right) = \frac{8\pi}{3} + 4\pi^2$$

378. (B) Each cross-section of the described solid is that of a washer, with area $A(x) = \pi(1+\tan x)^2 - \pi(1-\tan x)^2$. By symmetry about the x-axis, we see that $A(x) = 2[\pi(1+\tan x)^2 - \pi(1)^2] = 2\pi(\tan^2 x + 2\tan x$ as x varies from 0 to $\pi/4$. Therefore, the volume V of the solid is given by the integral:

$$V = 2\pi \int_0^{\pi/4} (\tan^2 x + 2\tan x)\,dx = 2\pi\left(\int_0^{\pi/4} \tan^2 x\,dx + 2\int_0^{\pi/4} \tan x\,dx\right) = 2\pi(I_1 + 2I_2)$$

To compute I_1, observe that $\tan^2 x = \sec^2 x - 1$. Therefore,

$$I_1 = \int_0^{\pi/4} \sec^2 x - 1\,dx = [\tan x - x]_0^{\pi/4} = 1 - \frac{\pi}{4}$$

For I_2, observe that $\tan x = \dfrac{\sin x}{\cos x}$. Therefore, we substitute $u = \cos x$, observe that $du = -\sin x\,dx$, and integrate with respect to u:

$$I_2 = \int_1^{1/\sqrt{2}} \frac{-du}{u} = [-\ln|u|]_1^{1/\sqrt{2}} = \frac{\ln 2}{2}$$

Combining these results yields our volume:

$$V = 2\pi(I_1 + 2I_2) = 2\pi\left(1 - \frac{\pi}{4} + \ln 2\right)$$

379. **(C)** The graphs in question are shown in the figure below.

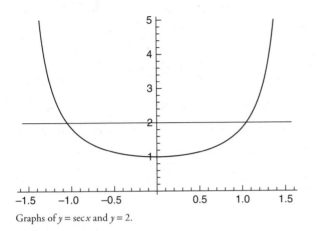

Graphs of $y = \sec x$ and $y = 2$.

Observe that each cross-section perpendicular to the x-axis is a washer having area $A(x) = 4\pi - \pi \sec^2 x$. To find the bounds of integration, we solve $f(x) = 2$ to obtain $x = \pm\dfrac{\pi}{3}$. Noting the symmetry about the y-axis, the volume of the solid is the value of the following integral:

$$2\pi \int_0^{\pi/3} 4 - \sec^2 x \, dx = 2\pi [4x - \tan x]_0^{\pi/3} = 2\pi \left(\frac{4\pi}{3} - \sqrt{3} \right)$$

380. **(D)** First, notice that the two curves intersect when x is a solution to $3 - x = \dfrac{2}{x}$. This equation reduces to $x^2 - 3x + 2 = 0$, which is only true if $x = 1$ or $x = 2$. These solutions are the bounds of the integral below.

Each cross-section is a washer with outer radius $R = 4 - x$ and inner radius $r = 1 + \dfrac{2}{x}$. Hence, the volume of the solid is computed thusly:

$$\pi \int_1^2 \left((4-x)^2 - \left(1 + \frac{2}{x}\right)^2 \right) dx = \pi \int_1^2 \left((4-x)^2 - 1 - \frac{4}{x} - \frac{4}{x^2} \right) dx$$

$$= \pi \left[-\frac{(4-x)^3}{3} - x - 4 \ln|x| + \frac{4}{x} \right]_1^2$$

$$= \pi \left[\left(-\frac{8}{3} - 4 \ln 2 \right) - (-9 - 1 + 4) \right] = \pi \left[\frac{10}{3} - 4 \ln 2 \right]$$

381. (B) The base is depicted below.

Observe that the hypotenuse of the right triangle may be represented by the equation $y = \frac{1}{3}x + 1$ from $x = 0$ to $x = 3$. Furthermore, observe this is also the side of each square cross-section at the point x. Therefore, the volume of S is equal to

$$\int_0^3 \left(\frac{1}{3}x + 1\right)^2 dx$$

In order to compute the integral, you set $u = \frac{1}{3}x + 1$, noting that under this substitution $3\,du = dx$, with u-bounds $u = 1$ and $u = 2$. Thus, your integral becomes

$$3\int_1^2 u^2\,du = 3\left[\frac{u^3}{3}\right]_1^2 = [u^3]_1^2 = 8 - 1 = 7$$

382. (E) Each cross-section perpendicular to the y-axis is a washer of area $A(y) = \pi(a^2 - y^2)$. The volume may be computed in terms of a:

$$\pi\int_0^a (a^2 - y^2)\,dy = \pi\left[a^2 y - \frac{y^3}{3}\right]_0^a = \pi\left(a^3 - \frac{a^3}{3}\right) = \frac{2\pi a^3}{3}$$

Thus, you solve $18\pi = \frac{2\pi a^3}{3}$ for a, to obtain $a = 3$.

383. (B) See the figure below.

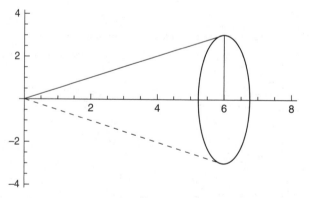

A cone generated by rotating $y = \frac{x}{2}$ about the x-axis.

An easy algebraic solution follows from the observation that the revolution of the triangle fills out a *cone* of height 6 and base radius 3. Hence,

$$V_{\text{Cone}} = \frac{1}{3}\pi \cdot 3^2 \cdot 6 = 18\pi$$

For a solution using calculus, note that the hypotenuse of the triangle is described by the equation $y = \dfrac{1}{2}x$ from $x = 0$ to $x = 6$. Therefore, the volume of the cone is obtained by the following computation:

$$\frac{\pi}{4}\int_0^6 x^2\,dx = \frac{\pi}{4}\left[\frac{x^3}{3}\right]_0^6 = \frac{6^3\,\pi}{12} = 18\pi$$

384. (C) Observe that $\dfrac{dx}{dt} = 4\sin t \cos t = 2\sin(2t)$ and $\dfrac{dy}{dt} = 2\cos(2t)$. Therefore, the surface area is given by the following integral computation:

$$\int_0^{\pi/2} 2\pi y \sqrt{\left(\frac{dx}{dt}\right)^2 + \left(\frac{dy}{dt}\right)^2}\,dt = \int_0^{\pi/2} 2\pi\sin(2t)\sqrt{4\sin^2(2t) + 4\cos^2(2t)}\,dt$$

$$= 4\pi\int_0^{\pi/2}\sin(2t)\,dt = 4\pi\left[-\frac{\cos(2t)}{2}\right]_0^{\pi/2} = 4\pi\left[\frac{1}{2} + \frac{1}{2}\right] = 4\pi$$

where you have used the Pythagorean identity $\sin^2\theta + \cos^2\theta = 1$ at the third step.

385. (B) Observe that $\dfrac{dx}{dt} = 2t$ and $\dfrac{dy}{dt} = 1$. Therefore, the surface area is given by the following integral computation:

$$\int_0^1 2\pi y \sqrt{\left(\frac{dx}{dt}\right)^2 + \left(\frac{dy}{dt}\right)^2}\,dt = 2\pi\int_0^1 t\sqrt{4t^2 + 1}\,dt$$

You make the substitution $u = 4t^2 + 1$, so that $du = 8t\,dt$ and the u-bounds run from $u = 1$ to $u = 5$. The integral thus becomes

$$\frac{\pi}{4}\int_1^5 \sqrt{u}\,du = \frac{\pi}{4}\left[\frac{1}{2\sqrt{u}}\right]_1^5 = \frac{\pi}{4}\left[\frac{1}{2\sqrt{5}} - \frac{1}{2}\right] = \frac{\pi}{8}\left(\frac{1-\sqrt{5}}{\sqrt{5}}\right).$$

386. (E) Since $\dfrac{dx}{dt} = 2e^{t/2}$ and $\dfrac{dy}{dt} = e^t - 4$, we see that surface area is given by

$$2\pi\int_0^1 (e^t - 4t)\sqrt{4e^t + (e^t - 4)^2}\,dt = 2\pi\int_0^1 (e^t - 4t)\sqrt{e^{2t} - 4e^t + 4}\,dt$$

$$= 2\pi\int_0^1 (e^t - 4t)(e^t - 2)\,dt = 2\pi\int_0^1 (e^{2t} - 4te^t - 2e^t + 8)\,dt$$

$$= 2\pi\left(\int_0^1 (e^{2t} - 2e^t + 8)\,dt - \int_0^1 4te^t\,dt\right) = 2\pi(I - J)$$

The integral I appearing above, may be computed as follows:

$$I = \int_0^1 (e^{2t} - 2e^t + 8)\,dt = \left[\frac{1}{2}e^{2t} - 2e^t + 8t\right]_0^1 = \left(\frac{e^2}{2} - 2e + 8 - \frac{1}{2} + 2\right) = \left(\frac{e^2}{2} - 2e + \frac{19}{2}\right)$$

To compute the integral J, we use by-parts integration. Let $u = t$ and $dv = e^t\,dt$ in the formula $\int u\,dv = uv - \int v\,du$. Then,

$$J = 4\int_0^1 te^t\,dt = 4\left([te^t]_0^1 - \int_0^1 e^t\,dt\right) = 4[te^t - e^t]_0^1 = 4(e^1 - e^1 + e^0) = 4$$

Thus, the surface area is

$$2\pi(I - J) = 2\pi\left(\frac{e^2}{2} - 2e + \frac{19}{2} - 4\right) = 2\pi\left(\frac{e^2}{2} - 2e + \frac{11}{2}\right) = \pi(e^2 - 4e + 11)$$

387. (D) Recall that the length of a curve with polar equation $r = f(\theta)$, $a \le \theta \le b$, is by

$$\int_a^b \sqrt{r^2 + \left(\frac{dr}{d\theta}\right)^2}\,d\theta \tag{1}$$

Therefore, the length of the curve $r = e^\theta$, $0 \le \theta \le 2\pi$, is

$$\int_0^{2\pi} \sqrt{r^2 + \left(\frac{dr}{d\theta}\right)^2}\,d\theta = \int_0^{2\pi} \sqrt{(e^\theta)^2 + (e^\theta)^2}\,d\theta$$

$$= \int_0^{2\pi} \sqrt{2(e^\theta)^2}\,d\theta = \sqrt{2}\int_0^{2\pi} e^\theta\,d\theta = \sqrt{2}[e^\theta]_0^{2\pi} = \sqrt{2}(e^{2\pi} - 1) \approx 755.885$$

388. (B) Observe that $r^2 = (\sin\theta - \cos\theta)^2 = \sin^2\theta - 2\sin\theta\cos\theta + \cos^2\theta = 1 - \sin(2\theta)$, and $\left(\dfrac{dr}{d\theta}\right)^2 = (\cos\theta + \sin\theta)^2 = 1 + \sin(2\theta)$. Therefore, the arc length is given by

$$\int_0^{\pi} \sqrt{r^2 + \left(\frac{dr}{d\theta}\right)^2}\,d\theta = \int_0^{\pi} \sqrt{2} = \sqrt{2}\pi$$

389. (C) You start by computing $\dfrac{dx}{dt} = -\sin t + \sin t + t\cos t = t\cos t$, and $\dfrac{dy}{dt} = \cos t - \cos t + t\sin t = t\sin t$. Next, you compute the following integral and set it equal to the length of the curve:

$$2\pi^2 = \int_0^a \sqrt{\left(\frac{dx}{dt}\right)^2 + \left(\frac{dy}{dt}\right)^2}\,dt = \int_0^a \sqrt{t^2\cos^2 t + t^2\sin^2 t}\,dt = \int_0^a t\,dt = \frac{a^2}{2}$$

Solving for a in the equation $2\pi^2 = \dfrac{a^2}{2}$ yields $a = 2\pi$.

390. (A) The fundamental theorem of calculus tells you that $F'(x) = f(x)$. Therefore, you solve $f(x) = 0$. Factoring yields $(x^2 - 1)(x^2 + 3) = 0$, which occurs at $x = \pm 1$. To use the second derivative test, you compute $F''(x) = f'(x) = 4t^3 + 4t = 4t(t^2 + 1)$. Note that $F''(-1) < 0$ and $F''(1) > 0$. Therefore, you conclude that $x = 1$ is the only value for which $F(x)$ has a local minimum.

For an area interpretation of this answer, observe that $F(x)$ is exactly the area underneath the curve $y = f(t)$ shown in the figure below, and note where $f(t)$ is above/below the x-axis.

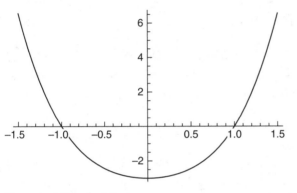

The graph of $f(t) = t^4 + 2t^2 - 3$.

391. (C) You break the interval $[0, \ln 8]$ into three even subintervals of length $\ln 2 = \dfrac{\ln 8}{3}$. The area under the curve is therefore approximated using 3 trapezoids by

$$\frac{\ln 8}{6}(e^0 + 2e^{\ln 2} + 2e^{2\ln 2} + e^{3\ln 2}) = \frac{\ln 2}{2}(1 + 4 + 8 + 8) = \frac{21\ln 2}{2}$$

392. (A) Observe that you may write the acceleration $a(t)$ as a vector sum of its $i = \langle 1, 0 \rangle$ component and its $j = \langle 0, 1 \rangle$ component. That is,

$$a(t) = \pi^2 \sin(\pi t) \cdot i + 6t \cdot j$$

The velocity is related to acceleration by $\int a(t)\,dt = v(t) + C$, for some constant vector C. Thus, the velocity of the particle at time t is

$$v(t) = \left(\pi^2 \int \sin(\pi t)\,dt\right) \cdot i + \left(\int 6t\,dt\right) \cdot j = -\pi \cos(\pi t) \cdot i + 3t^2 \cdot j + C$$

for a particular constant vector $C = \langle c_1, c_2 \rangle = c_1 \cdot i + c_2 \cdot j$ to be solved for later. **(Observe that when integrating vector functions you treat their i- and j-directional components separately.)**

The position of the particle at time t is obtained using the relation $\int v(t)\,dt = s(t) + D$ for a separate constant vector D. Thus, upon integrating once more, you obtain the position function at time t:

$$s(t) = \int v(t)\,dt = \int (-\pi \cos(\pi t) \cdot i + 3t^2 \cdot j + C)\,dt = -\sin(\pi t) \cdot i + t^3 \cdot j + Ct + D, \quad (2)$$

where you have once again gained a constant of integration, $D = d_1 \cdot i + d_2 \cdot j$.

You are given that $s(0) = O$, which may be written as $s(0) = \langle 0, 0 \rangle$, or $s(0) = 0 \cdot i + 0 \cdot j$. But, letting $t = 0$ in Equation (2) yields

$$s(0) = -\sin(\pi t) \cdot i + t^3 \cdot j + Ct + D = 0 \cdot i + 0 \cdot j + C(0) + D = O + D = D$$

Hence, $D = O$.

To solve for C, you use that $s(2) = i + j$. But, letting $t = 2$ in Equation (2) yields

$$s(2) = -\sin(2\pi) \cdot i + (2)^3 \cdot j + 2C = 0 \cdot i + 8 \cdot j + 2C = 8 \cdot j + 2C$$

Thus, you solve $i + j = 8 \cdot j + 2C$ to obtain: $2C = i - 7 \cdot j$. Division by 2 yields $C = \dfrac{1}{2}i - \dfrac{7}{2}j$. Therefore, using the above results and combining terms corresponding to their i-components and j-components, yields

$$s(t) = -\sin(\pi t) \cdot i + t^3 \cdot j + Ct + D = -\sin(\pi t) \cdot i + t^3 \cdot j + \left(\frac{1}{2}i - \frac{7}{2}j\right)t + O$$

$$= \left(\frac{1}{2} - \sin(\pi t)\right) \cdot i + \left(t^3 - \frac{7}{2}\right) \cdot j$$

393. (C) Recall that the length of a vector curve with equation $r(t) = \langle f(t), g(t) \rangle$, $a \le t \le b$, is

$$\int_a^b \|r'(t)\| dt \tag{3}$$

where $\|r'(t)\|$ is the magnitude of the vector function $r'(t) = \langle f'(t), g'(t) \rangle$. That is,

$$\|r'(t)\| = \sqrt{(f'(t))^2 + (g'(t))^2}$$

Observe that $\dfrac{d}{dt}(\ln|\sec t|) = \dfrac{1}{\sec t} \cdot (\sec t \tan t) = \tan t$. Also $\dfrac{d}{dt}(t) = 1$.

Then, you see that $r'(t) = \left\langle \dfrac{d}{dt}(\ln|\sec t|), \dfrac{d}{dt}(t) \right\rangle = \langle \tan t, 1 \rangle$.

Then, $\|r'(t)\| = \sqrt{\tan^2 t + 1} = \sqrt{\sec^2 t} = \sec t$, and the length of the curve is therefore

$$\int_0^{\pi/4} \|r'(t)\| dt = \int_0^{\pi/4} \sec t \, dt = \left[\ln|\sec t + \tan t|\right]_0^{\pi/4}$$

$$= \ln|\sec(\pi/4) + \tan(\pi/4)| - \ln|1 + 0| = \ln\left|\sqrt{2} + 1\right| - 0 = \ln\left|\sqrt{2} + 1\right| \approx 0.881$$

394. (E) Your plan is to use Equation (3). Thus, you differentiate $r(t) = \langle e^t, \sqrt{3}e^t \rangle$ separately along each directional component, to obtain: $r'(t) = \langle e^t, \sqrt{3}e^t \rangle$. Then,

$$\|r'(t)\| = \sqrt{(e^t)^2 + \left(\sqrt{3}e^t\right)^2} = \sqrt{(e^t)^2 + 3(e^t)^2} = \sqrt{4(e^t)^2} = 2e^t$$

You are given that the length of the curve is 8. Therefore, you use Equation (3) to obtain

$$8 = \int_0^k \|r'(t)\| \, dt = 2 \int_0^k e^t \, dt = 2(e^k - 1)$$

To solve for k, we take the equation $8 = 2(e^k - 1)$, divide by 2, and add 1 to obtain the reduced equation $5 = e^k$. Applying the natural logarithm to both sides yields $\ln 5 = \ln(e^k)$, which further reduces to $\ln 5 = k$.

395. (A) You are looking for solutions to $g'(x) = 0$, which by the fundamental theorem of calculus is equivalent to solving $f(x) = 0$. To begin, use properties of logarithms to simplify the equation:

$$\log_2(x - 1) + \log_2(x + 1) = 0 \Leftrightarrow \log_2[(x - 1)(x + 1)] = 0$$

Exponentiating both sides of the last equation base 2, this becomes

$$(x - 1)(x + 1) = 1 \Leftrightarrow x^2 - 2 = 0$$

You therefore obtain $x = \pm\sqrt{2}$ as possible solutions. Note, however, that $-\sqrt{2}$ cannot be a solution to this equation, in particular, $-\sqrt{2}$ is not in the domain of $\log_2(x - 1)$. Thus, the only x-value at which $y = g(x)$ has a horizontal tangent is when $x = \sqrt{2}$.

396. (A) $g(0) = \displaystyle\int_0^0 f(x) \, dx = 0.$

$$g(2) = \int_0^2 f(x) \, dx = \int_0^2 (4 - 4x) \, dx = [4x - 2x^2]_0^2 = 0.$$

$$g(6) = \int_0^6 f(x) \, dx = \int_0^2 (4 - 4x) \, dx + \int_2^3 (2x - 8) \, dx + \int_3^5 (4x - 14) \, dx + \int_5^6 6 \, dx$$

$$= 0 + (-3) + 4 + 6 = 7.$$

g is increasing on $[0, 1) \cup (\frac{7}{2}, 6]$, since $g'(t) = f(t) > 0$ on these intervals. Note that $\frac{7}{2}$ is the solution to $4x - 14 = 0$.

At $t = \frac{7}{2}$, g has a minimum value.

Since $g'(t) = f(t)$ is decreasing only on $(0, 2)$, you see that $g''(x) < 0$ on this interval. Therefore, g is concave down only on $(0, 2)$.

397. (A) Let A_i denote the area of R_i. Then,

$$A_1 = \int_0^{a^2} \sqrt{x}\, dx = \left[\frac{2}{3}x^{3/2}\right]_0^{a^2} = \frac{2a^3}{3}$$

and

$$A_2 = \int_0^a y^2\, dy = \left[\frac{y^3}{3}\right]_0^a = \frac{a^3}{3}$$

Therefore, you solve $A_1 = A_2$ for a:

$$\frac{2a^3}{3} = \frac{a^3}{3} \Leftrightarrow a^3 = 0 \Leftrightarrow a = 0$$

Thus, there are no solutions $a > 0$ such that $A_1 = A_2$.

(B) R_1 is divided into two regions of equal area, each expressed by the following integral equations:

$$\frac{A_1}{2} = \int_0^b \sqrt{x}\, dx = \frac{2}{3}b^{3/2}$$

and

$$\frac{A_1}{2} = \int_b^{a^2} \sqrt{x}\, dx = \frac{2}{3}a^3 - \frac{2}{3}b^{3/2}$$

Upon setting these two expressions equal, you solve for b:

$$\frac{2}{3}b^{3/2} = \frac{2}{3}a^3 - \frac{2}{3}b^{3/2} \Leftrightarrow 2b^{3/2} = a^3 \Leftrightarrow b = \frac{a^2}{\sqrt[3]{4}}$$

The solid obtained by rotating R_1 about the y-axis has cross-sections of washers, with inner radius $x = y^2$ and outer radius $x = a^2$. Therefore, the volume is expressed as

$$\pi \int_0^a (a^4 - y^4)\, dy = \pi \left[a^4 y - \frac{1}{5}y^5\right]_0^a = \pi\left(a^5 - \frac{1}{5}a^5\right) = \pi\left(\frac{5a^5}{5} - \frac{a^5}{5}\right) = \frac{4\pi a^5}{5}$$

At each y belonging to $[0, a]$, the area of a square cross-section is given by $A(y) = (y^2)^2 = y^4$. Therefore, the volume of the solid is just the value of the integral:

$$\int_0^a y^4\, dy = \frac{1}{5}a^5$$

398. (A) The line segment AB is defined by the equation $y = x + 1$ for $0 \leq x \leq 2$, the line segment BC is given by $y = -2x + 7$ for $2 \leq x \leq 3$, while AC is simply $y = 1$.

Note that each cross-section perpendicular to the x-axis is a washer with inner radius 1, outer radius $x + 1$ for $0 \leq x \leq 2$, and outer radius $7 - 2x$ for $2 \leq x \leq 3$. Therefore, the volume of the solid is given by the integral:

$$\pi \left(\int_0^2 [(x+1)^2 - 1^2] \, dx + \int_2^3 [(7-2x)^2 - 1^2] \, dx \right) = \pi \left(\left[\frac{(x+1)^3}{3} - x \right]_0^2 + \left[-\frac{(7-2x)^3}{6} - x \right]_2^3 \right)$$

$$= \pi \left(\left[(9-2) - \frac{1}{3} \right] + \left[\left(-\frac{1}{6} - 3 \right) + \left(\frac{27}{6} + 2 \right) \right] \right)$$

$$= \pi \left(\left[\frac{21}{3} - \frac{1}{3} \right] + \left[\left(-\frac{1}{6} - \frac{18}{6} \right) + \left(\frac{27}{6} + \frac{12}{6} \right) \right] \right)$$

$$= \pi \left(\frac{20}{3} + \frac{20}{6} \right) = \frac{30\pi}{3} = 10\pi$$

(B) Observe that AB may be written as $x = y - 1$ and BC as $x = \dfrac{7-y}{2}$ for $1 \leq y \leq 3$. Therefore, revolving R about the y-axis is the value of the following integral:

$$\pi \int_1^3 \left(\frac{7-y}{2} \right)^2 - (y-1)^2 \, dy = \pi \left(\frac{1}{4} \int_1^3 (7-y)^2 \, dy - \int_1^3 (y-1)^2 \, dy \right)$$

Next, you let $u = 7 - y$ in the first integral above, noting that under this substitution $-du = dy$. Therefore, the integral becomes

$$\pi \left(-\frac{1}{4} \int_{y=1}^{y=3} u^2 \, du - \int_1^3 (y-1)^2 \, dy \right) = \pi \left(-\frac{1}{4} \left[\frac{1}{3} u^3 \right]_{y=1}^{y=3} - \left[\frac{1}{3} (y-1)^3 \right]_1^3 \right)$$

$$= \pi \left(\frac{1}{4} \left[-\frac{1}{3} (7-y)^3 \right]_1^3 - \left[\frac{1}{3} (y-1)^3 \right]_1^3 \right)$$

$$= \pi \left(-\frac{1}{12} [4^3 - 6^3] - \frac{1}{3} [2^3] \right)$$

(*Note:* $4^3 - 6^3 = 4 \cdot 4^2 - 4 \cdot 9 \cdot 6$.) $\quad = \pi \left(-\frac{1}{12} [4(4^2 - 9 \cdot 6)] - \frac{8}{3} \right)$

$$= \pi \left(-\frac{1}{3} (16 - 54) - \frac{8}{3} \right)$$

$$= \pi \left(\frac{38}{3} - \frac{8}{3} \right) = 10\pi$$

(C) The area of each square cross-section from $x = 0$ to $x = 2$ is $A_1(x) = (x + 1 - 1)^2 = x^2$, while from $x = 2$ to $x = 3$, the area is given by $A_2(x) = (-2x + 7 - 1)^2 = (6 - 2x)^2 = 4(x^2 - 6x + 9)$. Therefore, the volume of the described solid is

$$\int_0^2 A_1(x)\,dx + \int_2^3 A_2(x)\,dx = \int_0^2 x^2\,dx + 4\int_2^3 (x^2 - 6x + 9)\,dx$$

$$= \left[\frac{1}{3}x^3\right]_0^3 + 4\left[\frac{1}{3}x^3 - 3x^2 + 9x\right]_2^3$$

$$= \left(\frac{8}{3}\right) + 4\left[(9 - 27 + 27) - \left(\frac{8}{3} - 12 + 18\right)\right]$$

$$= \left(4 \cdot \frac{2}{3}\right) + 4\left[9 - \frac{26}{3}\right]$$

$$= 4\left(9 - \frac{24}{3}\right) = 4[9 - 8] = 4$$

399. (A) The area bounded by the curves is given by

$$\int_0^{\pi/2} \cos x\,dx = [\sin x]_0^{\pi/2} = 1$$

(B) If you rotate the region about the x-axis, note that each cross-section perpendicular to the x-axis is a disc with radius $y = \cos x$. Therefore, you set up the following integral to compute the volume:

$$\pi \int_0^{\pi/2} \cos^2 x\,dx = \frac{\pi}{2}\int_0^{\pi/2}(1 + \cos(2x))\,dx = \frac{\pi}{2}\left[x + \frac{\sin(2x)}{2}\right]_0^{\pi/2} = \frac{\pi^2}{4}$$

(C) Cross-sections perpendicular to the x-axis are rectangles of depth $\sin x + 1$ and width $\cos x$. Therefore, the volume is given by

$$\int_0^{\pi/2} (\sin x + 1)\cos x\,dx$$

This integral is computable by many techniques. You will make the substitution $u = \sin x + 1$, so that $du = \cos x\,dx$, and the integral becomes

$$\int_1^2 u\,du = \frac{1}{2}[4 - 1] = \frac{3}{2} \text{ or } 1.5 \text{ ft}^3$$

400. (A) Observe that $\left(\dfrac{dx}{dt}\right)^2 = (\cos t + \sin t)^2 = \cos^2 t + 2\sin t \cos t + \sin^2 t = 1 + \sin(2t)$

and $\left(\dfrac{dy}{dt}\right)^2 = (\cos t - \sin t)^2 = 1 - \sin(2t)$. The length of the arc is therefore

$$\int_{\pi/4}^{\pi/2} \sqrt{\left(\frac{dx}{dt}\right)^2 + \left(\frac{dy}{dt}\right)^2}\ dt = \int_{\pi/4}^{\pi/2} \sqrt{(1 + \sin(2t)) + (1 - \sin(2t))}\ dt$$

$$= \int_{\pi/4}^{\pi/2} \sqrt{1 + 1}\ dt = \sqrt{2}\left(\frac{\pi}{2} - \frac{\pi}{4}\right) = \frac{\sqrt{2}\pi}{4}$$

(B) Since $x = \sin t - \cos t$, we observe that $x = 0$ when $t = \pi/4$ and $x = 1$ when $t = \pi/2$. You further observe that $dx = (\cos t + \sin t)\,dt$.

Hence, the area under the curve is the value of the following integral:

$$\int_0^1 y\,dx = \int_{\pi/4}^{\pi/2} (\sin t + \cos t)(\cos t + \sin t)\,dt = \int_{\pi/4}^{\pi/2} (\sin t + \cos t)^2\,dt$$

$$= \int_{\pi/4}^{\pi/2} \cos^2 t + 2\sin t \cos t + \sin^2 t\,dt = \int_{\pi/4}^{\pi/2} 1 + \sin(2t)\,dt$$

$$= \left[t + \frac{1}{2}\cos(2t)\right]_{\pi/4}^{\pi/2} = \left(\frac{\pi}{2} - \frac{1}{2}\right) - \left(\frac{\pi}{4} + 0\right) = \frac{2\pi}{4} - \frac{\pi}{4} - \frac{1}{2} = \frac{\pi}{4} - \frac{1}{2}$$

(C) Recall from the solution to part A that $\left(\dfrac{dx}{dt}\right)^2 = 1 + \sin(2t)$ and $\left(\dfrac{dy}{dt}\right)^2 = 1 - \sin(2t)$.

The surface area is given by the following integral:

$$2\pi \int_{\pi/4}^{\pi/2} y \sqrt{\left(\frac{dx}{dt}\right)^2 + \left(\frac{dy}{dt}\right)^2}\ dt = 2\pi \int_{\pi/4}^{\pi/2} (\sin t + \cos t)\sqrt{(1 + \sin(2t)) + (1 - \sin(2t))}\ dt$$

$$= 2\pi \int_{\pi/4}^{\pi/2} (\sin t + \cos t)\sqrt{1 + 1}\ dt$$

$$= 2\sqrt{2}\pi \int_{\pi/4}^{\pi/2} (\sin t + \cos t)\,dt$$

$$= 2\sqrt{2}\pi\left[-\cos t + \sin t\right]_{\pi/4}^{\pi/2}$$

$$= 2\sqrt{2}\pi(-\cos(\pi/2) + \sin(\pi/2) - (-\cos(\pi/4) + \sin(\pi/4)))$$

$$= 2\sqrt{2}\pi\left((0 + 1) - \left(-\frac{1}{\sqrt{2}} + \frac{1}{\sqrt{2}}\right)\right)$$

$$= 2\sqrt{2}\pi(0 + 1 - 0) = 2\sqrt{2}\pi$$

Chapter 9: More Applications of Definite Integrals

401. (D) The average value of a continuous function $f(x)$ on an interval $[a, b]$ is calculated with the formula $\dfrac{1}{b-a}\displaystyle\int_a^b f(x)\,dx$.

The integral of a graph is the area between the curve and the x-axis. You count area above the x-axis as positive area, and area below the x-axis as negative area.

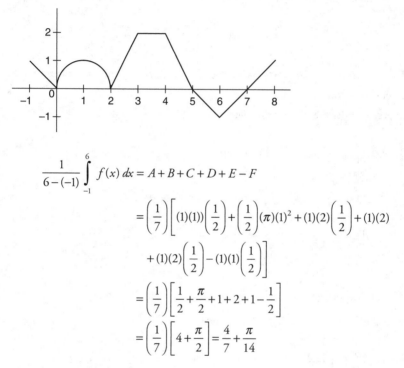

$$\frac{1}{6-(-1)}\int_{-1}^{6} f(x)\,dx = A + B + C + D + E - F$$

$$= \left(\frac{1}{7}\right)\left[(1)(1)\left(\frac{1}{2}\right) + \left(\frac{1}{2}\right)(\pi)(1)^2 + (1)(2)\left(\frac{1}{2}\right) + (1)(2)\right.$$

$$\left. + (1)(2)\left(\frac{1}{2}\right) - (1)(1)\left(\frac{1}{2}\right)\right]$$

$$= \left(\frac{1}{7}\right)\left[\frac{1}{2} + \frac{\pi}{2} + 1 + 2 + 1 - \frac{1}{2}\right]$$

$$= \left(\frac{1}{7}\right)\left[4 + \frac{\pi}{2}\right] = \frac{4}{7} + \frac{\pi}{14}$$

402. (E) The average value of a continuous function $f(x)$ on an interval $[a, b]$ is calculated with the formula $\dfrac{1}{b-a}\displaystyle\int_a^b f(x)\,dx$

$$\frac{1}{3-1}\int_1^3 (2t^2 - 14t - 5)\,dt = \frac{1}{2}\left[\frac{2t^3}{3} - \frac{14t^2}{2} - 5t\right]\Bigg|_1^3$$

$$= \left(\frac{1}{2}\right)\left(\left[\frac{2(3)^3}{3} - \frac{14(3)^2}{2} - 5(3)\right] - \left[\frac{2(1)^3}{3} - \frac{14(1)^2}{2} - 5(1)\right]\right)$$

$$= \left(\frac{1}{2}([-60] - \left[-\frac{34}{3}\right]\right) = -\frac{73}{3}$$

403. **(C)** The average value of a continuous function $f(x)$ on an interval $[a, b]$ is calculated with the formula $\dfrac{1}{b-a}\displaystyle\int_a^b f(x)\,dx$

$$\frac{1}{\frac{\pi}{6}-\left(-\frac{\pi}{2}\right)}\int_{-\frac{\pi}{2}}^{\frac{\pi}{6}}[3\sin(2x)-3\cos(2x)]\,dx$$

$$=\left(\frac{3}{2\pi}\right)(3)\left[-\frac{1}{2}\cos(2x)-\frac{1}{2}\sin(2x)\right]\Bigg|_{-\frac{\pi}{2}}^{\frac{\pi}{6}}$$

$$=-\frac{9}{2\pi}\left[\left[\cos\left(2\left(\frac{\pi}{6}\right)\right)+\sin\left(2\left(\frac{\pi}{6}\right)\right)\right]-\left[\cos\left(2\left(-\frac{\pi}{2}\right)\right)+\sin\left(2\left(-\frac{\pi}{2}\right)\right)\right]\right]$$

$$=-\frac{9}{2\pi}\left[\cos\left(\frac{\pi}{3}\right)+\sin\left(\frac{\pi}{3}\right)-\cos(-\pi)-\sin(-\pi)\right]$$

$$=-\frac{9}{2\pi}\left[\frac{1}{2}+\frac{\sqrt{3}}{2}-(-1)-0\right]=\frac{-27}{4\pi}-\frac{9\sqrt{3}}{4\pi}$$

404. **(A)** Recall that $a(t)=\displaystyle\int v(t)\,dt$ and integrals represent net change, so the area under the curve represents change in velocity. This area is positive for $[5, 15]$. To find the total change in velocity you calculate the area under the curve. Area above the x-axis is positive, and area below the x-axis is negative.

$$\text{Total change in velocity} = -(5)(10)\left(\frac{1}{2}\right)+(10)(5)\left(\frac{1}{2}\right)-(5)(5)\left(\frac{1}{2}\right)$$

$$=-\frac{25}{2}\ \text{m/s}$$

405. **(D)** Recall that the integral of velocity is the net change in distance. Graph the function in your calculator and find the first positive zero is at $t=2$. Use your calculator to find

$$\int_1^2 7\cos\left(\frac{1}{7}t^2+1\right)dt=1.635.$$

The change in distance between $t=1$ and $t=2$ is 1.635 m, and at $t=1$ the particle is at 10 m. Therefore the particle is at 11.635 m at $t=2$, when its velocity is 0 for the first time.

406. **(E)** The integral of acceleration is velocity. The area under the curve in the interval $[0,3]$ is triangular and below the x-axis:

$$-((3)(2)(.5))=-3\ \text{m/s}$$

Since speed is the absolute value of velocity, the correct answer is 3 m/s

407. **(B)** Total displacement $= \int_{0}^{15} v(t)\,dt$

$$= -(6)(5)\left(\frac{1}{2}\right) + (10)(10)\left(\frac{1}{2}\right) = 35$$

Total distance $= \int_{0}^{15} |v(t)|\,dt$

$$= \left| -(6)(5)\left(\frac{1}{2}\right) \right| + (10)(10)\left(\frac{1}{2}\right) = 65$$

Distance − displacement $= 65 - 35 = 30$ m

408. **(A)** Total displacement $= \int_{0}^{6} t^2 - t - 2\,dt$

$$= \frac{t^3}{3} - \frac{t^2}{2} - 2t \Big|_{0}^{6} = 42$$

To find total distance, find the zeros of the function and set up the integral piecewise.

$$t^2 - t - 2 = 0 \Rightarrow (t-2)(t+1) = 0 \Rightarrow t = 2, -1 \text{ are roots}$$

Only $t = 2$ is in the interval [0, 6]

Total distance $= \int_{0}^{2} |t^2 - t - 2|\,dt + \int_{2}^{6} |t^2 - t - 2|\,dt$

$$= \left| \frac{t^3}{3} - \frac{t^2}{2} - 2t \right|_{0}^{2} + \left| \frac{t^3}{3} - \frac{t^2}{2} - 2t \right|_{2}^{6}$$

$$= \left| -\frac{10}{3} \right| + \left| \frac{136}{3} \right|$$

$$= \frac{146}{3}$$

Distance − displacement $= \dfrac{146}{3} - 42 = \dfrac{20}{3}$

409. **(A)** Use your calculator to calculate the displacement integral.

Displacement $= \int_{0}^{5} t^3 - 3t^2 - 2t + 4\,dt = \dfrac{105}{4}$. Use your calculator to obtain the zeros of the function in the interval [0, 5].

$$\Rightarrow t \Rightarrow 1, 3.24$$

$$\text{Distance} = \left| \int_0^1 t^3 - 3t^2 - 2t + 4\, dt \right| + \left| \int_1^{3.24} t^3 - 3t^2 - 2t + 4\, dt \right| + \left| \int_{3.24}^5 t^3 - 3t^2 - 2t + 4\, dt \right|$$

$$= \left| \frac{9}{4} \right| + \left| -6.25 \right| + \left| 30.25 \right| = 38.75 \text{ m}$$

$$\text{Displacement} - \text{distance} = \frac{105}{4} - 38.75 = -12.5 \text{ m}$$

410. (E) $v(t) = \int a(t)\, dt = \int -3\, dt = -3t + C$

Since $v_0 = 9 \Rightarrow -3(0) + C = 9 \Rightarrow C = 9$

$$v(t) = -3t + 9$$

$$\text{Total distance} = \int_0^7 |-3t + 9|\, dt$$

To calculate this integral, find the zeros of the function and use them to write the integral piecewise.

$-3t + 9 = 0 \Rightarrow t = 3$ is a zero.

$$\int_0^7 |-3t + 9|\, dt = \left| \int_0^3 (-3t + 9)\, dt \right| + \left| \int_3^7 (-3t + 9)\, dt \right| = \frac{75}{2}$$

411. (B) The particle is moving to the right when the area under graph of the velocity function is above the x-axis. Use your calculator to find the zeros of the velocity function and evaluate the integral $\int_a^b v(t)\, dt$ when $[a, b]$ is an interval that bounds an area above the x-axis.

The zeros of $v(t) = 2 \sin(3t)$ in $[-\pi, \pi]$ are at $t = -3.14, -2.09, -1.05, 0, 1.05, 2.09, 3.14$.

$$\text{Distance} \approx \int_{-2.05}^{-1.05} 2 \sin(3t)\, dt + \int_0^{1.05} 2 \sin(3t)\, dt + \int_{2.05}^{3.14} 2 \sin(3t)\, dt \approx 4$$

412. (A) Marginal profit is the derivative of profit.

$$\text{Profit} = \int_0^{20000} (1000 - 0.04x)\, dx$$

$$= [1000x - 0.02x^2]\big|_0^{20000} = 12{,}000{,}000$$

413. (A) Marginal cost is the derivative of cost.

$$\text{Cost} = \int_0^{100} (5 + 0.4x)\, dx$$

$$= [5x + 0.2x^2]\big|_0^{100} = 2500$$

414. (D) Profit = revenue − cost

$$\text{Revenue} = \int (\text{marginal revenue})$$

$$\text{Cost} = \int (\text{marginal cost})$$

To find the revenue and the cost, calculate the area under the graphs of marginal revenue and marginal cost, respectively.

Profit = revenue − cost = $(1000)(50)(0.5) - (50)(50)(0.5) = 23750$

415. (C) Recall that integrals represent net change. The net change in the temperature of the penny can be represented by the integral:

$$\int_0^4 15e^{-.12t}\, dt = 47.65$$

The temperature of the penny has dropped by 47.65 degrees in 4 min. Recall that the penny started at a temperature of 150 degrees.

$150 - 47.65 = 102.35$

416. (B) To calculate the change in temperature during the day evaluate the integral:

$$\int_0^7 3\cos(t/3)\, dt = 6.5077$$

Since the temperature at 6 a.m. is 75 degrees, at 1 p.m. the temperature is

$75 + 6.51 = 81.51.$

417. (C) The area under the curve tells you the change in air temperature. Area below the x-axis represents a decrease in temperature, and area above the x-axis represents an increase in temperature.

$60 - (2)(2)(0.5) + (1)(1)(0.5) = 58.5$

418. (C) $\displaystyle\int_0^{12} 12e^{-.4t}\, dt = 29.7531$

419. (A) Since integral represents net change, the integral of the rate of leaking water is the net change in the amount of water leaked. The best approximation for the area under the curve from $t = 5$ to $t = 15$ is a triangle. Area of the triangle $= (10)(30)(0.5) = 150$

420. (A) The amount of water leaked after 12 h can be calculated by

$$\int_0^{12} 7e^{-.1t} + 5\, dt = 108.916.$$

Since the tank is half empty after 12 h, the total amount of water in the full tank was:

$(108.916)(2) = 217.833$

421. (D) Since $f(t)$ = the rate of increase of BBQ sales,

Let $F(x)$ = the increase in the number of BBQs sold after x weeks.

$$F(x) = \int_0^x f(t)\,dt$$

Since there were no BBQs sold in January,

Increase in number of BBQs sold = total number of BBQs sold

To find the number of BBQs sold after 7 weeks we calculate:

$$F(7) = \int_0^7 (30 + 20\ln(1+t))\,dt$$

Enter the integral in your calculator and obtain 402.711.

Thus Texas Depot will sell approximately 403 BBQs after the 7th week of February.

422. (E) Since $f(t)$ = the rate of increase of the bacteria population,

Let $F(x)$ = the increase in the bacteria population after x hours.

$$F(x) = \int_0^x f(t)\,dt$$

Thus, the population increase between the 6th and 8th hours is

$$F(8) - F(6) = \int_0^8 (100 + 250\ln(3+t))\,dt - \int_0^6 (100 + 250\ln(3+t))\,dt$$

$$= \int_6^8 (100 + 250\ln(3+t))\,dt$$

Enter the integral in your calculator to obtain 1350.46.

Thus, the population in the bacteria colony will increase by approximately 1350 between the 6th and 8th hours.

423. (C) Since the function in the graph represents the *rate of* increase of hybrid car production, the *integral* of this function is the increase of hybrid car production. On a graph,
you interpret the integral to mean area between the function and the *x*-axis. The integral
from 2008 to 2011 can best be approximated by a rectangle and triangle, as shown below.

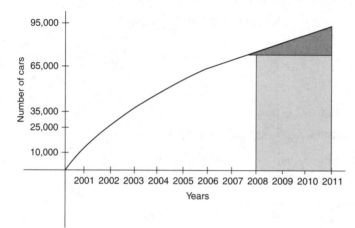

$$3(65,000) + 3(30,000)(0.5) = 240,000$$

So hybrid car production increased by approximately 240,000 cars between 2008 and 2011.

424. (A) This problem is asking us to apply the mean value theorem for integrals.

Mean value theorem for integrals

If $f(x)$ is continuous on $[a, b]$, then there exists a number c in $[a, b]$ such that

$$\int_a^b f(x)\,dx = f(c)(b - a)$$

First, you need to check the hypothesis of the mean value theorem. Since $f(x) =$
$\dfrac{x^2}{4} + 2$ is a polynomial, it is continuous and differential everywhere. So there is a time during the day where the price of the stock is equal to it's average price for the day.

Now you solve for "c" in the mean value theorem.

$$\int_0^6 \left(\frac{x^2}{4} + 2 \right) dx = f(c)(6 - 0)$$

$$\left. \frac{x^3}{12} + 2x \right|_0^6 = f(c)(6)$$

$$30 = f(c)(6) \quad 5 = f(c) \quad 5 = \frac{c^2}{4} + 2 \quad 3 = \frac{c^2}{4} \quad 12 = c^2$$

$$\pm 2\sqrt{12} = c$$
$$\pm 2\sqrt{3} = c$$

Since only $2\sqrt{3}$ is in the interval $[0, 6]$, $c = 2\sqrt{3}$

Thus approx. 3.4641 h after 9 a.m., the value of Abercrombie stock is equal to its average value for the day. Now convert 3.4641 h into hours and minutes.

$$(0.4641\ h)\left(\frac{60}{1}\ min\right) = 27.8486\ min$$

So approximately 3 h and 28 min after 9 a.m., the value of Abercrombie stock is equal to its average value for the day. The time is 12:28 p.m.

425. (C) The function $f(x) = \sqrt{2x - 3}$ is continuous for $x \geq \dfrac{3}{2}$, thus:

$$\int_{2}^{8} \sqrt{2x - 3}\ dx = f(c)(8 - 2)$$

$$\left. \frac{(2x - 3)^{3/2}}{3} \right|_{2}^{8} = 6f(c)$$

$$\left(\frac{1}{3}\right)\left((2(8) - 3)^{3/2} - (2(2) - 3)^{3/2}\right) = 6f(c)$$

$$15.2907 = 6f(c);\ \ f(c) = 2.54845;\ \ 2.54845 = \sqrt{2c - 3};\ \ c = 4.75$$

426. (B) Recall the equation for expotential growth/decay is

$$y = y_0 e^{kt}$$

Since the half-life is 8645 years and you start with 20 g

$$\Rightarrow 10 = 20e^{(k)(8645)}$$

$$\Rightarrow \frac{1}{2} = e^{8645k}$$

To solve the exponential equation for k you take the natural log of both sides.

$$\Rightarrow \ln\left(\frac{1}{2}\right) = \ln(e^{8645k})$$

$$\Rightarrow \ln(1) - \ln(2) = (8645k)(\ln(e))$$

Now recall that $\ln(1) = 0$ and $\ln(e) = 1$.

$$\Rightarrow -\ln(2) = 8645k \Rightarrow k = -\frac{\ln(2)}{8645} \Rightarrow y = 20e^{\frac{-\ln(2)}{8645}t}$$

Now to find how many grams are left after 1000 years, you can plug $t = 1000$ into the equation.

$$y = 20e^{\frac{-\ln(2)}{8645}(1000)} \approx 18$$

427. (A) Since the rate of increase is proportional to the amount of bacteria present, this is an exponential growth/decay model

$$y(t) = y_0 e^{kt}$$

$$y(1) = 200, \; y(3) = 600 \quad \Rightarrow \quad 200 = y_0 e^{k(1)}, \; 600 = y_0 e^{k(3)}$$

Since you have 2 equations and 1 unknown variable, y_0, you can use substitution to solve for y_0 and k.

Solving the first equation for y_0 you get: $y_0 = \dfrac{200}{e^k} \Rightarrow y_0 = 200e^{-k}$

Now substituting your value of y_0 into the second equation you get

$$600 = (200e^{-k})e^{3k}$$

$$600 = 200e^{2k}$$

$$3 = e^{2k}$$

$$\ln(3) = 2k$$

$$k = \frac{\ln(3)}{2}$$

$$k = \ln(\sqrt{3})$$

Now that you have k, you can plug back into your first equation to solve for y_0

$$k = \ln(\sqrt{3}), \; y_0 = 200e^{-k} \quad \Rightarrow \quad y_0 = 200e^{-\ln(\sqrt{3})} \quad \Rightarrow \quad y_0 \approx 115.47$$

To calculate how many bacteria are present after 1 week (7 days), you can use the following equation:

$$y(t) = 115.47e^{\ln(\sqrt{3})(t)}$$

$$y(7) = 115.47e^{(\ln(\sqrt{3}))(7)} \approx 5400$$

428. (D) Since you are given an equation where the rate of change of y is proportional to y, this is an exponential growth/decay model.

$$y(t) = y_0 e^{kt}$$

Since the temperature of the coffee decreases by 30% after 5 min:

$$\left(\frac{30}{100}\right) y_0 = y_0 e^{k(5)}$$

$$\frac{3}{10} = e^{5k}$$

$$\ln\left(\frac{3}{10}\right) = \ln(e^{5k})$$

$$\ln(3) - \ln(10) = 5k$$

$$k = \frac{\ln(3) - \ln(10)}{5}$$

429. (D) Since the rate of growth of the population is proportional to the population, this is an exponential growth/decay problem. Let $t =$ number of years after 2002.

$$y(t) = y_0 e^{kt}$$

Since the population increased by 11% between 2002 and 2011 (9 years) you may write

$$\left(y_0 + \frac{11}{100} y_0\right) = y_0 e^{k(9)}$$

$$\left(1 + \frac{11}{100}\right) y_0 = y_0 e^{k(9)}$$

$$(1.11) = e^{k(9)}$$

$$\ln(1.11) = \ln(e^{9k})$$

$$\ln(1.11) = 9k$$

$$k \approx 0.0116$$

430. (B) Remember that the derivative of a function represents the slope. Thus you may write

$$\frac{dy}{dx} = \frac{4xy}{2x^2 + 3}$$

You see that this is a seperable differential equation and you can solve for the the original function y by seperating out the variables onto opposite sides and then integrating.

$$\frac{dy}{y} = \frac{4x}{2x^2 + 3} dx$$

$$\int \frac{dy}{y} = \frac{4x}{2x^2 + 3} dx$$

The integral on the right-hand side can be solved by substitution using substitution

$$u = 2x^2 + 3 \text{ and } du = 4x \, dx.$$

$$\ln |y| = \ln(2x^2 + 3) + C_1$$

$$e^{\ln|y|} = e^{\ln(2x^2+3)+C_1}$$

$$|y| = e^{\ln(2x^2+3)} e^{C_1}$$

$$|y| = e^{C_1}(2x^2 + 3)$$

$$y = \pm e^{C_1}(2x^2 + 3)$$

C_1 can be solved for by using our intital information the the curve passes through the point (2, 11). So when $x = 2$ and $y = 11$ the above equation must hold.

$$11 = \pm e^{C_1}(2(2)^2 + 3)$$
$$11 = \pm e^{C_1}(11)$$
$$1 = \pm e^{C_1} \Rightarrow C_1 = 0 \quad \text{and} \quad y = (2x^2 + 3)$$

Check the solution by differentiating.

431. (E) To solve the differential equation, seperate the variables out on two sides and integrate to solve for y.

$$\frac{dy}{dx} = -x\cos(x^2)$$
$$dy = -x\cos(x^2)\,dx$$
$$\int dy = \int -x\cos(x^2)\,dx$$

To solve the integral on the right use substitution with $u = x^2$ and $du = 2x\,dx$

$$y = -\int \frac{du}{2}\cos(u)$$
$$y = -\frac{1}{2}\int \cos(u)\,du$$
$$y = -\frac{1}{2}\sin(u) + C$$
$$y = -\frac{1}{2}\sin(x^2) + C$$

C can be solved for using our initial information that at $x = 0$, $y = 2$.

$$2 = -\frac{1}{2}\sin(0) + C \Rightarrow C = 2$$
$$y = -\frac{1}{2}\sin(x^2) + 2$$

You can check your solution by differentiating.

432. (C) To find the correct value of y, first solve the seperable differential equation.

$$\frac{dy}{dx} = 5x^4 y^2$$
$$\frac{1}{y^2}\,dy = x^4\,dx$$

$$\int \frac{1}{y^2} dy = \int x^4 dx$$

$$\frac{-1}{y} = \frac{5x^5}{5} + C$$

$$y = \frac{-1}{x^5 + C}$$

To solve for C, use the initial conditions that $y = 1$ when $x = 1$.

$$1 = \frac{-1}{1+C} \Rightarrow 1 + C = -1 \Rightarrow C = -2$$

$$y = \frac{-1}{x^5 - 2}$$

Now substitute $x = -1$ into the equation to find the value of y.

$$y = \frac{-1}{(-1)^5 - 2} = \frac{-1}{-1 - 2} = \frac{1}{3}$$

433. (A) Rewrite $\dfrac{d^2}{dx^2}$ as $\dfrac{dy'}{dx}$.

$$\frac{dy'}{dx} = 4x - 5$$

$$dy' = (4x - 5) dx \Rightarrow \int dy' = \int (4x - 5) dx \Rightarrow y' = 2x^2 - 5x + C_1$$

Solve for C_1 by using the initial condition that at $x = 0$, $y' = 3$.

$$3 = 2(0)^2 - 5(0) + C_1 \Rightarrow C_1 = 3$$

$$y' = 2x^2 - 5x + 3; \ y' = \frac{dy}{dx}; \ \frac{dy}{dx} = 2x^2 - 5x + 3$$

$$dy = (2x^2 - 5x + 3) dx \Rightarrow \int dy = \int (2x^2 - 5x + 3) dx$$

$$y = \frac{2}{3}x^3 - \frac{5}{2}x^2 + 3x + C_2$$

Solve for C_2 by using the initial condition that at $x = 0$, $y = 4$.

$$4 = \frac{2}{3}(0)^3 - \frac{5}{2}(0)^2 + 3(0) + C_2 \Rightarrow C_2 = 4$$

$$y = \frac{2}{3}x^3 - \frac{5}{2}x^2 + 3x + 4$$

Verify the solution by differentiating.

434. (B) By observing different properties of the slope field, we can eliminate the answer choices. Notice that in the column $x = 0$, the slope is constantly 0. So when $x = 0$, the slope is independent of the changing y-value in that column. This eliminates the choices (A) $\dfrac{dy}{dx} = y - x$ and (E) $\dfrac{dy}{dx} = y - x$.

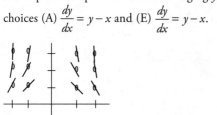

Now notice that when x and y are positive, the slope in the graph is negative. This eliminates choices (C) and (D). The correct answer is (B) $\dfrac{dy}{dx} = -xy$

435. (A) If you look vertically at any column of tangents, you'll notice that the tangents have the same slope. (Points in the same column have the same x-coordinate but different y-coordinates). Therefore, the numerical value of $\dfrac{dy}{dx}$ (which represents the slope of the tangent) depends only the x-coordinate of the point and is independent of the y-coordinate. Only choices (A) and (C) satisfy these conditions. Also notice that the tangents have positive slope when $x \geq 0$ and negative slope when $x \leq 0$.

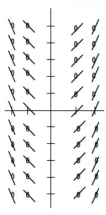

Therefore, the correct choice is (A), $\dfrac{dy}{dx} = x$.

436. (C) To find the family of functions whose slope field is given by $\dfrac{dy}{dx} = \dfrac{x}{y}$, solve the diffential equation by separating the variables and integrating both sides.

$$\frac{dy}{dx} = \frac{x}{y} \Rightarrow y\,dy = x\,dx \Rightarrow \int y\,dy = \int x\,dx$$

$$\frac{y^2}{2} = \frac{x^2}{2} + c_1 \Rightarrow -\frac{x^2}{2} + \frac{y^2}{2} = c_1$$

Given an initial condition, you could have found a particular solution by solving for c_1. This is equation is one of a hyperbola.

437. (B) To identify the slope field, make a table of values and sketch the slope field for yourself.

	$x = -2$	$x = -1$	$x = 0$	$x = 1$	$x = 2$
$y = 2$	0	−1	−2	−3	−4
$y = 1$	1	0	−1	−2	−3
$y = 0$	2	1	0	−1	−2
$y = -1$	3	2	1	0	−1
$y = -2$	4	3	2	1	0

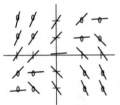

438. (D) Recall the following equations for logistic models:

$$\frac{dP}{dt} = kP\left(1 - \frac{P}{K}\right), \quad P(t) = \frac{K}{Ae^{-kt} + 1}, \quad A = \frac{K - P_0}{P_0}$$

Let $t =$ the number of years after 2001. So $t = 0$ represents 2001, and $t = 7$ represents 2008. From the problem you can fill in $K = 75$, $P_0 = 17$, $P(0) = 17$, $P(7) = 35$.

$$A = \frac{K - P_0}{P_0} \Rightarrow A = \frac{75 - 17}{17} \approx 3.41176$$

$$P(t) = \frac{K}{Ae^{-kt} + 1} = \frac{75}{3.41176e^{-kt} + 1}$$

$$P(7) = 35 \Rightarrow 35 = \frac{75}{3.41176e^{-k7} + 1}$$

$$(3.41176)e^{-7k} + 1 = \frac{75}{35} \approx 2.14286$$

$$e^{-7k} \approx .334976$$

$$\ln(e^{-7k}) \approx \ln(.334976) \Rightarrow -7k \approx -1.0937 \Rightarrow k \approx 0.2$$

The year 2013 corresponds to $t = 12$.

$$P(12) = \frac{75}{3.41176e^{-(0.2)(12)} + 1} \approx 57$$

So in 2013, the rabbit population is approximately 57.

439. (C) Recall the following equations for logistic models:

$$\frac{dP}{dt} = kP\left(1 - \frac{P}{K}\right), \; P(t) = \frac{K}{Ae^{-kt} + 1}, \; A = \frac{K - P_0}{P_0}$$

From the information in the question, you see that $k = .71$, $K = 60$, $P_0 = 1$.

$$A = \frac{K - P_0}{P_0} = \frac{60 - 1}{1} = 59$$

$$P(t) = \frac{K}{Ae^{-kt} + 1} = \frac{60}{59e^{-.71t} + 1}$$

To determine the date when half the trees (i.e., 30 trees) will be infected you solve

$$30 = \frac{60}{59e^{-.71t} + 1} \Rightarrow 59e^{-.71t} + 1 = 2$$

$$e^{-.71t} = 0.0169 \Rightarrow \ln(e^{-.71t}) = \ln(0.0169) \Rightarrow -.71t = -4.08 \Rightarrow t \approx 6$$

After 6 days, half the trees will be infected.

440. (A) Recall the following equations for logistic models:

$$\frac{dP}{dt} = kP\left(1 - \frac{P}{K}\right), \; P(t) = \frac{K}{Ae^{-kt} + 1}, \; A = \frac{K - P_0}{P_0}$$

Let t = the number of days after Monday. So $t = 0$ means Monday, $t = 1$ means Tuesdays, etc. From information in the question, you see that $K = 1532$, $P_0 = 7$, $P(1) = 84$.

$$A = \frac{K - P_0}{P_0} = \frac{1532 - 7}{7} = 217.857$$

$$P(t) = \frac{K}{Ae^{-kt} + 1} = \frac{1532}{217.857e^{-kt} + 1}$$

Now by using $P(1) = 84$ you can solve for k.

$$84 = \frac{1532}{217.857e^{-k(1)} + 1} \Rightarrow 217.857e^{-k} + 1 = \frac{1532}{84} \Rightarrow e^{-k} \approx 0.079$$

$$\ln(e^{-k}) = \ln(0.079) \Rightarrow -k = -2.53831$$

Now the equation for the logistic model is

$$P(t) = \frac{1532}{217.857e^{-2.5383t} + 1}$$

Since Thursday refers to $t = 3$, to predict how many students have heard the rumor by Thursday you evaluate:

$$P(3) = \frac{1532}{217.857e^{-2.5383(3)} + 1} \approx 1384$$

So by Thursday approximately 1384 students have heard the rumor.

441. (D) Recall the following equations for logistic models:

$$\frac{dP}{dt} = kP\left(1 - \frac{P}{K}\right), \quad P(t) = \frac{K}{Ae^{-kt} + 1}, \quad A = \frac{K - P_0}{P_0}$$

Let $t =$ the number of days after the first diagnosis. From the problem you see that $K = 300$, $P(0) = 4$, $P(7) = 17$.

$$A = \frac{K - P_0}{P_0} = \frac{300 - 4}{4} = 74$$

$$P(t) = \frac{K}{Ae^{-kt} + 1} = \frac{300}{74e^{-kt} + 1}$$

You can use the information that $P(7) = 17$ to solve for k.

$$17 = \frac{300}{74e^{-k(7)} + 1} \Rightarrow 74e^{-7k} + 1 = \frac{300}{17} \Rightarrow e^{-7k} \approx 0.225 \Rightarrow \ln(e^{-7k}) = \ln(0.225)$$
$$-7k \approx -1.49 \Rightarrow k \approx 0.213$$

To estimate the number of students infected after 2 weeks, you evaluate

$$P(14) = \frac{300}{74e^{-(0.213)(14)} + 1} \approx 63.1$$

Thus after 2 weeks, approximately 64 of the quarantined students are infected with measles.

442. (E) The problem requires us to use Euler's method. First, calculate the step size.

$$\Delta t = \frac{3 - 1}{5} = 0.4$$

Recall that the general equation for calculating approximations with Euler's method is

$$y_n = y_{n-1} + \Delta x \cdot f'(x_{n-1}, y_{n-1})$$

Now construct a table to insert the approximations at each step.

t	y_n	$2\sin(4\pi t)$	$y_n = y_{n-1} + \Delta x \cdot f'(x_{n-1}, y_{n-1})$
1	2	0	$y_1 = 2 + 0.4(0) = 2$
1.4	2	−1.9	$y_2 = 2 + 0.4(-1.9) = 1.24$
1.8	1.24	−1.18	$y_3 = 1.24 + 0.4(-1.18) = 0.768$
2.2	0.768	1.18	$y_4 = 0.768 + 0.4(1.18) = 1.24$
2.6	1.24	1.9	$y_5 = 1.24 + 0.4(1.902) = 2$
3	2		

$y(3) \approx 2$

443. (B) Recall that the general equation for calculating approximations with Euler's method is

$$y_n = y_{n-1} + \Delta x \cdot f'(x_{n-1}, y_{n-1})$$

For the case of the the evaluation, transform $\dfrac{dy}{dx} + 4x^3 y = 2x^3$ to $\dfrac{dy}{dx} = 2x^3(1 - 2y)$.
Now construct a table to insert the approximations at each step.

x	y_n	$2x^3(1-2y)$	$y_n = y_{n-1} + \Delta x \cdot f'(x_{n-1}, y_{n-1})$
0	2	0	$y_1 = 2 + 0.2(0) = 20$
0.2	2	−0.048	$y_2 = 2 + 0.2(-0.048) = 1.9904$
0.4	1.9904	−0.382	$y_3 = 1.9904 + 0.2(-0.382) = 1.914$
0.6	1.914	−1.2217	$y_4 = 1.914 + 0.2(-1.2217) = 1.6697$
0.8	1.6697	−2.396	$y_5 = 1.6697 + 0.2(-2.396) = 1.19$
1	1.19		

$y(1) \approx 1.19$

444. (A) The problem requires us to use Euler's method. First, calculate the step size.

$$\Delta t = \frac{4-2}{3} = 0.67$$

Recall that the general equation for calculating approximations with Euler's method is

$$y_n = y_{n-1} + \Delta x \cdot f'(x_{n-1}, y_{n-1})$$

Now construct a table to insert the approximations at each step.

t	P_n	$-.2P\left(1-\dfrac{3P}{5}\right)$	$y_n = y_{n-1} + \Delta x \cdot f'(x_{n-1}, y_{n-1})$
2	3	0.48	$y_1 = 3 + 0.67(0.48) = 3.3216$
2.67	3.3216	0.6596	$y_2 = 3.3216 + 0.67(0.6596) = 3.764$
3.3	3.764	0.947	$y_3 = 3.764 + 0.67(0.947) = 4.398$
4	4.398		

$y(4) \approx 4.4$

445. (C) Recall that the general equation for calculating approximations with Euler's method is

$$y_n = y_{n-1} + \Delta x \cdot f'(x_{n-1}, y_{n-1})$$

Also recall that the velocity function shown is the derivative of the distance function, and that the ball begins 50 ft above the ground. Now construct a table and use the graph to fill in information as needed.

x	y_n	f(x)	$y_n = y_{n-1} + \Delta x \cdot f'(x_{n-1}, y_{n-1})$
0	50	0	$y_1 = 50 + 0.5(0) = 50$
0.5	50	-5	$y_2 = 50 + 0.5(-5) = 47.5$
1	47.5	-7	$y_3 = 47.5 + 0.5(-7) = 44$
1.5	44	-11	$y_4 = 44 + 0.5(-11) = 38.5$
2	38.5		

After 2 s, the ball is approximately 38.5 ft above the ground.

446. (A) $F(0) = 105$, $f(t)$ represents the rate of change of the temperature, 2 a.m. is 7 h after 7 p.m.

$$F(7) = 105 + \int_0^7 (-3 \sin(t/3))\, dt \Rightarrow F(7) = 105 + (9 \cos(t/3) \mid_0^7 \Rightarrow F(7) \approx 89.78$$

(B) $F(t) = 105 + \int_0^t (-3 \sin(x/3))\, dx$

(C) Average value $= \dfrac{1}{12 - 0} \int_0^{12} (-3 \sin(t/3))\, dt \approx -1.24°F$

(D) Since $f(t)$ is continuous on $[0, 12]$, by the mean value theorem for integrals there exists a number c in $[0, 12]$ such that

$$\int_0^{12} f(t)\,dt = f(c)(b - a)$$

You can use our solution from part D to show that $f(c) = -1.24$. Now solve for c.

$$f(c) = -1.24 \Rightarrow -3\sin(t/3) = -1.24 \Rightarrow \sin(t/3) = 0.413 \; t/3 = \sin^{-1}(0.413) \Rightarrow t \approx 0.426$$

Approximately 0.426 h after 7 p.m., the rate of change of the temperature in the greenhouse equals the average rate of change of the temperature for the night. Now convert 0.426 to minutes to find the time.

$$(0.426 \text{ h}) \times (60) = 25.56 \text{ min} \approx 26 \text{ min}$$

7:26 p.m.

447. (A) Set up a table of values for $\dfrac{dy}{dx} = \dfrac{4xy}{5}$ at the 15 given points.

	x = -2	x = -1	x = 0	x = 1	x = 2
y = 1	$-\dfrac{8}{5}$	$-\dfrac{4}{5}$	0	$\dfrac{4}{5}$	$\dfrac{8}{5}$
y = 2	$-\dfrac{16}{5}$	$-\dfrac{8}{5}$	0	$\dfrac{8}{5}$	$\dfrac{16}{5}$
y = 3	$-\dfrac{25}{5}$	$-\dfrac{12}{5}$	0	$\dfrac{12}{5}$	$\dfrac{24}{5}$

Then sketch the tangents at the various points as shown in the figure below.

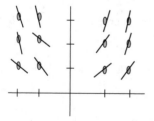

(B) Recall that the general equation for calculating approximations with Euler's method is

$$y_n = y_{n-1} + \Delta x \cdot f'(x_{n-1}, y_{n-1})$$

Now construct a table to insert the approximations at each step.

x	y_n	$\dfrac{4xy}{5}$	$y_n = y_{n-1} + \Delta x \cdot f'(x_{n-1}, y_{n-1})$
0	5	0	$y_1 = 5 + 0.1(0) = 5$
0.1	5	0.4	$y_2 = 5 + 0.1(0.4) = 5.04$
0.2	5.04	0.8064	$y_3 = 5.04 + 0.1(0.8064) = 5.1206$
0.3	5.1206		

$f(0.3) \approx 5.1206$

    (C) Solve the seperable differential equation.

$$\frac{dy}{dx} = \frac{4xy}{5} \Rightarrow \frac{1}{y}\,dy = \frac{4}{5}x\,dx$$

$$\int \frac{1}{y}\,dy = \int \frac{4}{5}x\,dx \Rightarrow \ln|y| = \frac{2}{5}x^2 + c_1$$

$$e^{\ln|y|} = e^{\frac{2}{5}x^2 + c_1} = e^{\frac{2}{5}x^2} \cdot e^{c_1}$$

$$y = c_2 e^{\frac{2}{5}x^2}$$

Using the initial condition $f(0) = 5$, you can solve for c_2.

$$5 = c_2 e^{\frac{2}{5}0^2} \Rightarrow c_2 = 5$$

    (D) Now use the work above to evaluate $y(0.3)$.

$$y(0.3) = 5e^{\frac{2}{5}(0.3)^2} = 5.8133$$

448.   **(A)** The slope at $x = 3$ (i.e., the point $(3, 4)$) is

$$\frac{dy}{dx} = \frac{2y}{3x^2} \Rightarrow \frac{dy}{dx} = \frac{2(4)}{3(3)^2} = \frac{8}{27}$$

Now write the equation of the tangent line at $(3, 4)$ in point-slope form.

$$y - y_0 = m(x - x_0) \Rightarrow y - 4 = \frac{8}{27}(x - 3) \Rightarrow y = \frac{8}{27}(x - 3) + 4$$

(B) $f(5) = \dfrac{8}{27}(5-3) + 4 = \dfrac{124}{27}$

(C) Start by seperating the variables and intergrating both sides

$$\frac{dy}{dx} = \frac{2y}{3x^2} \Rightarrow \frac{1}{2y}dy = \frac{1}{3x^2}dx \Rightarrow \int \frac{1}{2y}dy = \int \frac{1}{3x^2}dx$$

$$\frac{1}{2}\int \frac{1}{y}dy = \frac{1}{3}\int x^{-2}dx \Rightarrow \frac{1}{2}\ln|y| = \left(\frac{1}{3}\right)\frac{x^{-1}}{-1} + c_1 \Rightarrow \frac{1}{2}\ln|y| = -\frac{1}{3x} + c_1$$

$$\ln|y| = -\frac{2}{3x} + c_2 \Rightarrow e^{\ln|y|} = e^{-\frac{2}{3x}+c_2} = e^{-\frac{2}{3x}} \cdot e^{c_2}$$

$$y = c_3 e^{-\frac{2}{3x}}$$

Now use the initial condition $f(3) = 4$ to solve for c_3.

$$4 = c_3 e^{-\frac{2}{3(3)}} \Rightarrow c_3 = 4.9954$$

$$y = 4.9954 e^{-\frac{2}{3x}}$$

(D) Evaluate the equation in C at $f(5)$.

$$f(5) = 4.9954 e^{-\frac{2}{3(5)}} \approx 4.37184$$

449. (A) The amount of water leaked out after 7 min:

$$\int_0^7 15\sin\left(\frac{\pi t}{30}\right)dt = \frac{-15\cos\left(\dfrac{\pi t}{30}\right)\Big|^7}{\pi/30}\Bigg|_0 = -\frac{450}{\pi}\cos\left(\frac{\pi t}{30}\right)\Bigg|_0^7 \approx 36.7918$$

After 7 min, approximately 37 gal have leaked out of the water barrel.

(B) The average amount of water leaked out per minute from $t=0$ to $t=7$:

$$\frac{1}{7-0}\int_0^7 15\sin\left(\frac{\pi t}{30}\right)dt \approx \frac{1}{7}(36.7918) \approx 5.255 \approx 5 \text{ gal}$$

(C) The amount of oil in the drum at time t:

$$f(t) = 200 - \int_0^t 15\sin\left(\frac{\pi x}{30}\right)dx$$

(D) Let a be the value of t and solve the following equation for a:

$$50 = 200 - \int_0^a 15 \sin\left(\frac{\pi t}{30}\right) dt$$

$$50 = 200 - \left(-\frac{450}{\pi} \cos\left(\frac{\pi t}{30}\right)\bigg|_0^a\right)$$

$$50 = 200 - \left(-\frac{450}{\pi} \cos\left(\frac{\pi a}{30}\right) - -\frac{450}{\pi} \cos\left(\frac{\pi(0)}{30}\right)\right)$$

$$50 = 200 + \frac{450}{\pi} \cos\left(\frac{\pi a}{30}\right) - \frac{450}{\pi} \Rightarrow \cos\left(\frac{\pi a}{30}\right) \approx -0.0472$$

$$\frac{\pi a}{30} \approx \cos^{-1}(-0.0472) \approx 1.618 \Rightarrow a \approx 6.18$$

After approximately 6 min, there will be 50 gal left in the water barrel.

450. (A) Seperate the variables of the differential equation and integrate both sides.

$$\frac{dP}{dt} = \frac{1}{7}P\left(1 - \frac{P}{21}\right) \Rightarrow \frac{1}{P\left(1 - \dfrac{P}{21}\right)} dP = \frac{1}{7} dt \Rightarrow \frac{21}{P(21-P)} dP = \frac{1}{7} dt$$

$$\frac{dP}{P} + \frac{dP}{21-P} = \frac{1}{7} dt \Rightarrow \int \frac{dP}{P} + \int \frac{dP}{21-P} = \int \frac{1}{7} dt \Rightarrow \ln|P| - \ln|21-P| = \frac{t}{7} + c_1$$

$$\ln\left|\frac{P}{21-P}\right| = \frac{t}{7} + c_1$$

Exponentiate both sides, then isolate P.

$$\frac{P}{21-P} = e^{\frac{t}{7}} \cdot c_2 \Rightarrow P = \frac{21c_2 e^{\frac{t}{7}}}{1 + c_2 e^{\frac{t}{7}}}$$

Use both possible initial conditions to solve for c_2 and evaluate the limits.

$$P(0) = 15 \Rightarrow 15 = \frac{21c_2}{1 + c_2} \Rightarrow 21c_2 = 15(1 + c_2) \Rightarrow 6c_2 = 15 \Rightarrow c_2 = 2.5$$

$$P(t) = \frac{21(2.5)e^{\frac{t}{7}}}{1 + 2.5e^{\frac{t}{7}}} \Rightarrow P(t) = \frac{52.5e^{\frac{t}{7}}}{1 + 2.5e^{\frac{t}{7}}}$$

To calculate the limit as t tends to infinity, first factor out $e^{t/7}$ from the numerator and denominator.

$$P(t) = \frac{e^{t/7}(52.5)}{e^{t/7}(e^{-t/7} + 2.5)} = \frac{52.5}{(e^{-t/7} + 2.5)}$$

$$\lim_{x \to \infty} \frac{52.5}{e^{-t/7} + 2.5} = \frac{52.5}{2.5} = 21$$

(B) The rate of growth of the population is given by

$$\frac{dP}{dt} = \frac{1}{7}P\left(1 - \frac{P}{21}\right)$$

To find the maximum value of this function, calculate the derivative, set it equal to zero, and solve.

$$\frac{d^2P}{dt^2} = \frac{1}{7}P\left(-\frac{1}{21}\right) + \frac{1}{7}\left(1 - \frac{P}{21}\right) = \frac{1}{7} - \frac{2P}{147}$$

$$\frac{1}{7} - \frac{2P}{147} = 0 \Rightarrow P = 10.5$$

So the maximum of $\dfrac{dP}{dt}$ occurs when $P = 10.5$. Evaluate $\dfrac{dP}{dt}$ at $P = 10.5$ to find the maximum value.

$$\frac{dP}{dt} = \frac{1}{7}P\left(1 - \frac{P}{21}\right) \Rightarrow \frac{dP}{dt} = \frac{1}{7}(10.5)\left(1 - \frac{10.5}{21}\right) \Rightarrow \frac{dP}{dt} = 0.75$$

The population is growing fastest, at a rate of 75%, when the population is 10.5.

(C) Seperate the variables and integrate both sides.

$$\frac{dQ}{dt} = \frac{1}{7}Q\left(1 - \frac{t}{21}\right) \Rightarrow \frac{1}{Q}dQ = 17\left(\frac{21-t}{21}\right)dt \Rightarrow \int\frac{1}{Q}dQ = \frac{1}{147}\int(21-t)\,dt$$

$$\ln|Q| = \frac{1}{147}\left(21t - \frac{t^2}{2}\right) + c_1 \Rightarrow \ln|Q| = \frac{t}{7} - \frac{t^2}{294} + c_1$$

Now exponentiate both sides to solve for Q.

$$Q = c_2 e^{(t/7 - t^2/294)}$$

Use the initial condition, $Q(0) = 10$, to solve for c_2.

$$10 = c_2 e^0 \Rightarrow c_2 = 10 \Rightarrow Q(t) = 10e^{(t/7 - t^2/294)}$$

(D) For the function found in part (C), as t increases notice that

$$\frac{t}{7} - \frac{t^2}{294} \to -\infty \text{ so } e^{(t/7 - t^2/294)} \to 0$$

Therefore $\lim_{t \to \infty} Q(t) = 0$.

Chapter 10: Series (for Calculus BC Students Only)

451. (B) Observe that $a_n = \dfrac{n}{2n^2-3} = \dfrac{n}{n\left(2n-\dfrac{3}{n}\right)} = \dfrac{1}{2n-\dfrac{3}{n}} \to 0$ as $n \to \infty$. Therefore,

$\{a_n\}$ converges to 0. However, notice that for $n \geq 2$, we have $a_n = \dfrac{n}{2n^2-3} > \dfrac{n}{2n^2} = \dfrac{1}{2n}$. Therefore,

$$\sum_{n=2}^{\infty} a_n > \frac{1}{2} \sum_{n=2}^{\infty} \frac{1}{n} = \infty$$

and we have shown $\sum_{n=1}^{\infty} a_n = a_1 + \sum_{n=2}^{\infty} a_n$ diverges.

452. (C) Observe that $a_5 = (a_1 + a_2 + a_3 + a_4 + a_5) - (a_1 + a_2 + a_3 + a_4) = s_5 - s_4$. Therefore,

$$a_5 = s_5 - s_4 = \frac{3(5)+1}{2(5)-5} - \frac{3(4)+1}{2(4)-5} = \frac{16}{5} - \frac{13}{3} = \frac{16\cdot3 - 13\cdot5}{5\cdot3} = \frac{48-65}{15} = -\frac{17}{15}$$

453. (B) Observe that $\sum_{i=1}^{\infty} a_i = \lim_{n\to\infty} s_n$. Thus, we make the following computation:

$$\lim_{n\to\infty} \frac{(\ln n)^2}{n} = \lim_{n\to\infty} \frac{2\ln n \cdot \dfrac{1}{n}}{1} = 2\lim_{n\to\infty} \frac{\ln n}{n} = 2\lim_{n\to\infty} \frac{\left(\dfrac{1}{n}\right)}{1} = 0$$

where we have made two applications of l'Hospital's rule. Therefore, the series $\sum_{i=1}^{\infty} a_i$ converges to 0.

454. (D) This is a geometric series, as each term differs from the previous term by a factor of $\dfrac{1}{3}$. Observe that if you let $a_n = \dfrac{4}{3^{n-1}}$, then $a_1 = 4$, $a_2 = \dfrac{4}{3}$, $a_3 = \dfrac{4}{9}$, and so on. Thus, the series may be written

$$\sum_{n=1}^{\infty} a_n = 4 \sum_{n=1}^{\infty} \left(\frac{1}{3}\right)^{n-1} = \frac{4}{1-\left(\dfrac{1}{3}\right)} = \frac{4}{\left(\dfrac{2}{3}\right)} = 4\cdot\frac{3}{2} = 6$$

455. (E) Observe that $1.3\overline{12} = 1.3 + 0.012 + 0.00012 + 0.0000012 + \cdots = \dfrac{13}{10} + \dfrac{12}{10^3} +$

$\dfrac{12}{10^5} + \dfrac{12}{10^7} = \dfrac{13}{10} + 12 \sum_{n=1}^{\infty} \left(\dfrac{1}{10}\right)^{2n+1} = \dfrac{13}{10} + \dfrac{12}{10} \sum_{n=1}^{\infty} \left(\dfrac{1}{10^2}\right)^n$.

Therefore, you have clearly written the number as involving a geometric series of the form $a \sum_{n=1}^{\infty} r^n$, with $a = \dfrac{12}{10}$ and $r = \dfrac{1}{100}$.

You compute the sum of the geometric series:

$$a \sum_{n=1}^{\infty} r^n = ar \sum_{n=1}^{\infty} r^{n-1} = \frac{ar}{1-r} = \frac{\left(\dfrac{12}{10}\right)\left(\dfrac{1}{100}\right)}{1 - \dfrac{1}{100}} = \left(\frac{12}{1000}\right)\left(\frac{100}{99}\right) = \frac{12}{990}.$$

Putting this all together, you have

$$1.3\overline{12} = \frac{13}{10} + \frac{12}{990} = \frac{13 \cdot 99 + 12}{990} = \frac{13 \cdot (100-1) + 12}{990} = \frac{1300 - 13 + 12}{990} = \frac{1299}{990} = \frac{433}{330}.$$

456. (E) Observe that $3^{2n} 2^{3-3n} = 9^n \dfrac{2^3}{2^{3n}} = 8\left(\dfrac{9}{8}\right)^n$. Therefore, $s = 8 \sum_{i=1}^{\infty} \left(\dfrac{9}{8}\right)^n$ diverges, since this is a geometric series with $|r| = \left|\dfrac{9}{8}\right| > 1$.

457. (A) Observe that the nth partial sum may be computed, as follows:

$$s_n = \left(\frac{1}{1} - \frac{1}{3}\right) + \left(\frac{1}{2} - \frac{1}{4}\right) + \left(\frac{1}{3} - \frac{1}{5}\right) + \left(\frac{1}{4} + \frac{1}{6}\right) + \cdots + \left(\frac{1}{n-2} - \frac{1}{n}\right)$$

$$+ \left(\frac{1}{n-1} - \frac{1}{n+1}\right) + \left(\frac{1}{n} - \frac{1}{n+2}\right) = 1 + \frac{1}{2} - \frac{1}{n+1} - \frac{1}{n+2}$$

Therefore, the infinite series sums to

$$\lim_{n \to \infty} s_n = \lim_{n \to \infty}\left(1 + \frac{1}{2} - \frac{1}{n+1} - \frac{1}{n+2}\right) = 1 + \frac{1}{2} = \frac{3}{2}$$

458. (A) You apply basic algebraic properties of convergent series:

$$\sum_{n=1}^{n} (6a_n - 2b_n) = 6 \sum_{n=1}^{n} a_n - 2 \sum_{n=1}^{n} b_n = 6 \cdot 3 - 2 \cdot 7 = 4$$

459. (D) Consider the improper integral

$$\int_{2}^{\infty} \frac{1}{x(\ln x)^c} \, dx$$

You make the substitution $u = \ln x$, so that $du = \dfrac{1}{x} \, dx$, and the integral becomes

$$\lim_{t \to \infty} \int_{\ln 2}^{t} u^{-c} \, du$$

If $c = 1$, then

$$\lim_{t \to \infty} \int_{\ln 2}^{t} u^{-c}\,du = \lim_{t \to \infty} \int_{\ln 2}^{t} \frac{1}{u}\,du = \lim_{t \to \infty} (\ln t - \ln(\ln 2)) = \infty$$

If $c \neq 1$, then

$$\lim_{t \to \infty} \int_{\ln 2}^{t} u^{-c}\,du = \lim_{t \to \infty} \left[\frac{u^{1-c}}{1-c} \right]_{\ln 2}^{t} = \lim_{t \to \infty} \frac{t^{1-c}}{1-c} - \frac{(\ln 2)^{1-c}}{1-c}$$

Observe that $\lim_{t \to \infty} \dfrac{t^{1-c}}{1-c} = \infty$ when $c < 1$. However, $\lim_{t \to \infty} \dfrac{t^{1-c}}{1-c} = 0$ when $c > 1$.

Therefore, by the integral test, $\displaystyle\sum_{n=1}^{\infty} \dfrac{1}{n(\ln n)^c}$ is convergent when $c > 1$.

460. (A) You use the limit comparison test, comparing $a_n = \dfrac{5}{n^2 - n}$ with $b_n = \dfrac{5}{n^2}$. Observe that

$$\frac{\left(\dfrac{5}{n^2 - n}\right)}{\left(\dfrac{5}{n^2}\right)} = \frac{n^2}{n^2 - n} = \frac{n^2}{n^2\left(1 - \dfrac{1}{n}\right)} = \frac{1}{1 - \dfrac{1}{n}} \to 1$$

as $n \to \infty$. Therefore, $\displaystyle\sum_{i=2}^{\infty} \dfrac{5}{n^2 - n}$ converges, since it is clear that $\displaystyle\sum_{i=2}^{\infty} \dfrac{5}{n^2}$ converges.

461. (B) Observe that as $n \to \infty$, you have $\dfrac{3n+7}{5n-2} \to \dfrac{3}{5} \neq 0$. Therefore, the series diverges (by the divergence test).

462. (A) Consider the improper integral $\displaystyle\int_{1}^{\infty} \dfrac{\ln x}{x^2}\,dx$. You integrate by parts, letting $u = \ln x$ and $dv = \dfrac{dx}{x^2}$ in the formula $\displaystyle\int u\,dv = uv - \int v\,du$, so that

$$\lim_{t \to \infty} \int_{1}^{t} \frac{\ln x}{x^2}\,dx = \lim_{t \to \infty} \left[-\frac{\ln x}{x} \right]_{1}^{t} + \int_{1}^{t} \frac{1}{x^2}\,dx = \lim_{t \to \infty} \left(-\frac{\ln t}{t} \right) - \left[\frac{1}{x} \right]_{1}^{t}$$

$$= \lim_{t \to \infty} \left(-\frac{\ln t}{t} - \frac{1}{t} + 1 \right)$$

By l'Hospital's rule, $\lim_{t \to \infty} \dfrac{\ln t}{t} = \lim_{t \to \infty} \dfrac{\left(\dfrac{1}{t}\right)}{1} = 0$. Therefore,

$$\lim_{t \to \infty} \left(-\frac{\ln t}{t} - \frac{1}{t} + 1 \right) = -0 - 0 + 1 = 1 < \infty$$

The integral test then says that the series $\displaystyle\sum_{i=1}^{\infty} \dfrac{\ln n}{n^2}$ converges.

463. (A) Let $a_n = \dfrac{2 + \sin n}{n^2}$, and consider $b_n = \dfrac{3}{n^2}$. Note that $-1 \le \sin n \le 1$, so that $|a_n| \le b_n$, and also note $\sum_{n=1}^{\infty} b_n$ converges by the p-test. Therefore, $\sum_{n=1}^{\infty} a_n$ converges by comparison.

464. (A) Consider the limit $\lim_{n \to \infty} \dfrac{\sin\left(\dfrac{1}{n^2}\right)}{\dfrac{1}{n^2}}$, and note that by l'Hospital's rule

$$\lim_{n \to \infty} \frac{\sin\left(\dfrac{1}{n^2}\right)}{\dfrac{1}{n^2}} = \lim_{n \to \infty} \frac{\cos\left(\dfrac{1}{n^2}\right) \cdot (\cancel{-2n^{-3}})}{\cancel{2n^{-3}}} = \cos(0) = 1$$

Thus, as $\sum_{n=1}^{\infty}$ converges, $\sum_{n=1}^{\infty} \sin\left(\dfrac{1}{n^2}\right)$ converges by limit comparison.

465. (B) Notice that $\dfrac{n^n}{n!} > 1$, as is seen upon expanding and grouping as shown:

$$\frac{n^n}{n!} = \frac{n \cdot n \cdot n \cdots n \cdot n}{n \cdot (n-1) \cdot (n-2) \cdots 2 \cdot 1} = \frac{n}{n} \cdot \frac{n}{n-1} \cdot \frac{n}{n-2} \cdots \frac{n}{2} \cdot \frac{n}{1} > 1 \cdot 1 \cdot 1 \cdots 1 \cdot 1 = 1$$

Therefore, $\sum_{n=1}^{\infty} \dfrac{n^n}{n!} \ge \sum_{n=1}^{\infty} 1 = \infty$, and you have shown $\sum_{n=1}^{\infty} \dfrac{n^n}{n!}$ diverges by comparison.

466. (A) Notice $0 < \dfrac{1}{\ln(n+1)} < \dfrac{1}{\ln n}$, and moreover $\dfrac{1}{\ln n} \to 0$ as $n \to \infty$. Then, as $\sum_{n=2}^{\infty} (-1)^n \dfrac{1}{\ln n}$ is an alternating series, it therefore converges by the alternating series test.

467. (A) $\sum_{n=1}^{\infty} (-1)^{n-1} \dfrac{e^{1/n}}{\sqrt{n}}$ is an alternating series. Let $b_n = \dfrac{e^{1/n}}{\sqrt{n}}$. Observe that

$$\lim_{n \to \infty} b_n = \lim_{n \to \infty} \frac{e^{1/n}}{\sqrt{n}} = 0$$

since $e^{1/n} \to 1$ and $\sqrt{n} \to \infty$ as $n \to \infty$. Next, we show b_n is a decreasing sequence by considering $f(x) = \dfrac{e^{1/x}}{\sqrt{x}}$. Observe that

$$f'(x) = \frac{\sqrt{x} e^{1/x} \left(-\dfrac{1}{x^2}\right) - e^{1/x}\left(\dfrac{1}{2\sqrt{x}}\right)}{x} = \frac{-e^{1/x}\left(\dfrac{1}{x^{3/2}} + \dfrac{1}{2\sqrt{x}}\right)}{x} < 0$$

for all $x \ge 1$. Therefore, $f(n) = b_n$ is a decreasing sequence for $n \ge 1$. Thus, you apply the alternating series test to conclude that $\sum_{n=2}^{\infty} (-1)^{n-1} \dfrac{e^{1/n}}{\sqrt{n}}$ converges.

468. **(C)** Observe that $\cos(\pi n) = (-1)^n$. Hence $(-1)^n \dfrac{\cos(\pi n)}{n} = (-1)^{2n}\dfrac{1}{n} = \dfrac{1}{n}$, and

$$\sum_{n=1}^{\infty} (-1)^n \frac{\cos(\pi n)}{n} = \sum_{n=1}^{\infty} \frac{1}{n} = \infty.$$

Therefore, the series is divergent.

469. **(A)** Consider the ratio $\left|\dfrac{a_{n+1}}{a_n}\right| = \dfrac{\sin n + n}{2n+1}$ for $n \geq 1$. Then, as $-1 \leq \sin n \leq 1$, you see

that $\left|\dfrac{a_{n+1}}{a_n}\right| \leq \dfrac{1+n}{2n+1} \to \dfrac{1}{2}$ as $n \to \infty$. Therefore, $\displaystyle\sum_{n=0}^{\infty} a_n$ converges by the ratio test.

470. **(A)** Consider the integral $\displaystyle\int_1^{\infty} x^2 e^{-2x}\,dx$. You integrate by parts, letting $u = x^2$, and

$dv = e^{-2x}\,dx$ in the formula $\displaystyle\int u\,dv = uv - \int v\,du$, so that

$$\lim_{t\to\infty} \int_1^t x^2 e^{-2x}\,dx = \lim_{t\to\infty}\left[-\frac{1}{2}x^2 e^{-2x}\right]_1^t + \frac{1}{2}\int_1^t xe^{-2x}\,dx. \tag{1}$$

Notice that $\displaystyle\lim_{t\to\infty}\frac{t^2}{2e^{2t}} = \lim_{t\to\infty}\frac{2t}{4e^{2t}} = \lim_{t\to\infty}\frac{2}{8e^{2t}} = 0$, where you have applied l'Hospital's
rule twice. For the remaining integral, we again integrate by parts, choosing $w = x$, and
$dz = e^{-2x}\,dx$ so that

$$\lim_{t\to\infty}\int_1^t w\,dz = wz - \int z\,dw = \lim_{t\to\infty}\left[-\frac{1}{2}xe^{-2x}\right]_1^t + \frac{1}{2}\int_1^t e^{-2x}\,dx$$

$$= \lim_{t\to\infty}\left[-\frac{1}{2}xe^{-2x}\right]_1^t - \left[\frac{1}{4}e^{-2x}\right]_1^t$$

Then, as in Eq. (1), $\displaystyle\lim_{t\to\infty}\frac{1}{2}te^{-2t} \to 0$. Further, it is clear that $\displaystyle\lim_{t\to\infty}e^{-2t} \to 0$.
Therefore, as these are the only terms involving t, the improper integral is convergent.
Returning to the series under consideration, the integral test now applies to show that
$\displaystyle\sum_{n=0}^{\infty} n^2 e^{-2n}$ converges.

471. **(B)** The given series $\dfrac{1}{2} - \dfrac{2}{5} + \dfrac{3}{10} - \dfrac{4}{17} + \cdots = \displaystyle\sum_{n=1}^{\infty} (-1)^{n-1}\dfrac{n}{n^2+1}$. Since $\dfrac{n}{n^2+1} \to 0$

as $n \to \infty$. Furthermore, if you consider $f(x) = \dfrac{x}{x^2+1}$, and note that $f'(x) =$

$\dfrac{(x^2+1)-2x^2}{(x^2+1)^2} = \dfrac{1-x^2}{(x^2+1)^2} < 0$ for all $x > 1$, then you see that $f(n) = \dfrac{n}{n^2+1}$ is a decreasing

sequence. By the alternating series test, this series converges.

Next, consider $\left|(-1)^{n-1}\dfrac{n}{n^2+1}\right| = \dfrac{n}{n^2+1}$, and observe that

$$\left(\frac{n}{n^2+1}\right)\left(\frac{n}{1}\right) = \frac{n^2}{n^2+1} \to 1 \qquad \text{as} \qquad n \to \infty$$

Then, by limit comparison with $\dfrac{1}{n}$, you see that your series does not converge absolutely. Therefore, your series converges conditionally.

472. (A) Let $a_n = (-1)^{n+1}\left(\dfrac{n}{2n+1}\right)^n$, and observe that $|a_n| = \left(\dfrac{n}{2n+1}\right)^n$. You consider the nth root

$$\sqrt[n]{|a_n|} = \dfrac{n}{2n+1}$$

and observe that $\sqrt[n]{|a_n|} \to \dfrac{1}{2} < 1$ as $n \to \infty$. Therefore, by the root test, $\displaystyle\sum_{n=1}^{\infty} (-1)^{n+1}\left(\dfrac{n}{2n+1}\right)^n$ converges absolutely.

473. (C) You first show that the series does indeed converge. Note that $\dfrac{(-2)^n n}{8^n} = \dfrac{(-1)^n 2^n n}{8^n} = \dfrac{(-1)^n n}{4^n}$.

Next, you observe the series $\displaystyle\sum_{n=1}^{\infty} \dfrac{(-1)^n n}{4^n}$ is alternating. Then, using l'Hospital's rule you see that

$$\lim_{n\to\infty} \dfrac{n}{4^n} = \lim_{n\to\infty} \dfrac{1}{4^n \ln 4} = 0$$

Furthermore, if you consider $f(x) = \dfrac{x}{4^x}$, then

$$f'(x) = \dfrac{4^x - x4^x \ln 4}{4^{2x}} = \dfrac{4^x(1 - x\ln 4)}{4^{2x}} < 0 \qquad \text{for all} \qquad x \geq 1$$

Therefore, $f(n) = \dfrac{n}{4^n}$ is a decreasing sequence. Hence, the series converges by the alternating series test.

Now, let $b_n = \dfrac{n}{4^n}$, let $s = \displaystyle\sum_{n=1}^{\infty} (-1)^n b_n$, and let s_m denote the mth partial sum. Then, since you are working with a convergent alternating series:

$$|s - s_m| \leq |s_{m+1} - s_m| = b_{m+1} = \dfrac{(m+1)}{4^{m+1}}$$

Notice that

$$b_8 = \dfrac{1}{8192} \approx 0.00012207$$

and

$$s_7 = \left(-\dfrac{1}{4} + \dfrac{2}{4^2} - \dfrac{3}{4^3} + \dfrac{4}{4^4} - \dfrac{5}{4^5} + \dfrac{6}{4^6} - \dfrac{7}{4^7}\right) \approx -0.160095215$$

Therefore,

$$|s - s_7| \leq b_8$$

This error of 0.00012 does not affect the third decimal place, so $s \approx -0.160$ is correct to three decimal places.

474. (E) Let $a_n = \dfrac{x^n}{n2^n}$. Then, if $x \neq 0$, we have

$$\left| \frac{a_{n+1}}{a_n} \right| = \left| \frac{x^{n+1}}{(n+1)2^{n+1}} \cdot \frac{n2^n}{x^n} \right| = \frac{n}{2(n+1)} |x| \to \frac{|x|}{2} \quad \text{as} \quad n \to \infty$$

By the ratio test, the series is absolutely convergent, hence convergent, when $\dfrac{|x|}{2} < 1$, and divergent when $\dfrac{|x|}{2} > 1$. That is, the series converges when $|x| < 2$ the diverges when $|x| > 2$. The ratio test gives no information when $|x| = 2$. Therefore, you must check the cases $x = \pm 2$ separately.

When $x = 2$, you are working with the harmonic series $\displaystyle\sum_{n=1}^{\infty} \frac{1}{n}$, which diverges.

When $x = -2$, you have the alternating series $\displaystyle\sum_{n=1}^{\infty} \frac{(-1)^n}{n}$. This is convergent by the alternating

series test, since $\dfrac{1}{n+1} < \dfrac{1}{n}$ and $\lim_{n\to\infty} \dfrac{1}{n} \to 0$.

Therefore, $\displaystyle\sum_{n=1}^{\infty} \frac{x^n}{n2^n}$ is convergent only when $x \in [-2, 2)$.

475. (B) Let $a_n = n^3(x-4)^n$. Then, if $x \neq 4$, we have

$$\left| \frac{a_{n+1}}{a_n} \right| = \left| \frac{(n+1)^3(x-4)^{n+1}}{n^3(x-4)^n} \right| = \left(\frac{n+1}{n} \right)^3 |x-4| \to |x-4| \quad \text{as} \quad n \to \infty$$

By the ratio test, the series is absolutely convergent, hence convergent, when $|x-4| < 1$, and divergent when $|x-4| > 1$. The radius of convergence, therefore, is 1.

476. (C) By the root test, an infinite series Σa_n is absolutely convergent, hence convergent, when $\lim_{n\to\infty} \sqrt[n]{|a_n|}$ and divergent when $\lim_{n\to\infty} \sqrt[n]{|a_n|} > 1$.

Let $a_n = \dfrac{x^{2n}}{(\ln n)^n}$. Then, since

$$\lim_{n\to\infty} \sqrt[n]{|a_n|} = \lim_{n\to\infty} \sqrt[n]{\left| \frac{x^{2n}}{(\ln n)^n} \right|} = \lim_{n\to\infty} \left| \frac{x^2}{\ln n} \right| = 0 < 1$$

for all x, you see that the interval of convergence is $(-\infty, \infty)$.

477. (E) Observe that

$$\frac{x}{2+3x} = \left(\frac{x}{2} \right)\left(\frac{1}{1 + \left(\dfrac{3x}{2} \right)} \right) = \left(\frac{x}{2} \right)\left(\frac{1}{1 - \left(\dfrac{-3x}{2} \right)} \right)$$

and recall the geometric series formula for $|x| < 1$:

$$\frac{1}{1-x} = \sum_{n=0}^{\infty} x^n$$

You replace x by $-\dfrac{3x}{2}$ in the formula, yielding

$$\frac{x}{2+3x} = \left(\frac{x}{2}\right)\left(\frac{1}{1-\left(\frac{-3x}{2}\right)}\right) = \left(\frac{x}{2}\right)\sum_{n=0}^{\infty}\left(\frac{-3x}{2}\right)^n$$

You will simplify this series further in a moment. Stopping at this step makes it simpler to find the interval of convergence. Now, since this is a geometric series, it converges if and only if $\left|\dfrac{-3x}{2}\right| < 1$. That is, it converges when $|x| < \dfrac{2}{3}$ and diverges when $|x| \ge \dfrac{2}{3}$.

Therefore, the interval of convergence is $\left(-\dfrac{2}{3}, \dfrac{2}{3}\right)$.

You now continue the simplification of your series:

$$\left(\frac{x}{2}\right)\sum_{n=0}^{\infty}\left(\frac{-3x}{2}\right)^n = \left(\frac{x}{2}\right)\sum_{n=0}^{\infty}\frac{(-3)^n}{2^n}x^n = \sum_{n=0}^{\infty}\frac{(-3)^n}{2^{n+1}}x^{n+1} = \sum_{n=1}^{\infty}\frac{(-3)^{n-1}}{2^n}x^n$$

Hence, $\dfrac{x}{2+3x} = \displaystyle\sum_{n=1}^{\infty}\frac{(-3)^{n-1}}{2^n}x^n$.

478. (A) Recall the geometric series formula for $|x| < 1$:

$$\frac{1}{1-x} = \sum_{n=0}^{\infty} x^n$$

You are given that $f(x) = \ln(1-2x)$. Observe that $f'(x) = \dfrac{-2}{1-2x}$. Replacing x by $2x$ in the above formula, you find

$$f'(x) = \frac{-2}{1-2x} = -2\sum_{n=0}^{\infty}(2x)^n = -2\sum_{n=0}^{\infty}2^n x^n = -\sum_{n=0}^{\infty}2^{n+1}x^n$$

As this is a geometric series, it converges if and only if $|2x| < 1$. That is, it converges when $|x| < \dfrac{1}{2}$ and diverges when $|x| \ge \dfrac{1}{2}$. Therefore, its radius of convergence is $r = \dfrac{1}{2}$.

Note that you have found only a power series representation for $f'(x)$. In order to obtain a power series representation for $f(x) = \ln(1-2x)$ then, you must integrate. The caution here is that, in doing so, we obtain a constant of integration, as shown below:

$$-\int\left(\sum_{n=0}^{\infty}2^{n+1}x^n\right)dx = -\sum_{n=0}^{\infty}2^{n+1}\left(\int x^n dx\right) = -\left(\sum_{n=0}^{\infty}2^{n+1}\frac{x^{n+1}}{n+1}\right)+c = -\left(\sum_{n=1}^{\infty}\frac{2^n}{n}x^n\right)+c$$

To solve for the constant of integration, c, notice that

$$f(0) = \ln(1 - 2(0)) = \ln 1 = 0$$

When you substitute 0 in for x in the series above, you obtain

$$-\sum_{n=1}^{\infty} \frac{2^n}{n}(0)^n + c = c$$

Combining these two observations, you find that $c = 0$, and you conclude

$$\ln(1 - 2x) = -\sum_{n=1}^{\infty} \frac{2^n}{n}x^n$$

with the same radius of convergence $r = \dfrac{1}{2}$.

For a more rigorous explanation of why the radius of convergence of $f(x)$ is the same as the radius of convergence for $f'(x)$, you may use the root test. Let $a_n = \dfrac{2^n}{n}x^n$, and assume for the moment that you know $\lim_{n \to \infty} n^{1/n} = 1$. (You will prove this fact later.) Then

$$\lim_{n \to \infty} \sqrt[n]{|a_n|} = \lim_{n \to \infty} \sqrt[n]{\left|\frac{2^n}{n}x^n\right|} = \lim_{n \to \infty} \left|\frac{2}{n^{1/n}}x\right| = \left|\frac{2}{1}x\right| = 2|x|$$

And, this quantity is less than 1 if and only if $|x| < \dfrac{1}{2}$. Therefore, the radius of convergence of the series is $r = \dfrac{1}{2}$.

To complete your solution, you fill in the missing step above, showing $\lim_{n \to \infty} n^{1/n} = 1$. In computing $\lim_{n \to \infty} n^{1/n}$, you first observe that it has the indeterminate form ∞^0. You rewrite the expression and use l'Hospital's rule to evaluate the limit:

$$\lim_{n \to \infty} n^{1/n} = \lim_{n \to \infty} e^{\ln(n^{1/n})} = \lim_{n \to \infty} e^{\frac{1}{n}\ln n} = \lim_{n \to \infty} e^{\frac{\ln n}{n}} = \lim_{n \to \infty} e^{\frac{\left(\frac{1}{n}\right)}{1}} = \lim_{n \to \infty} e^{\frac{1}{n}} = e^0 = 1$$

where you note that l'Hospital's rule was used at the fourth step above. Therefore, you have shown $\lim_{n \to \infty} n^{1/n} = 1$, and this completes our solution.

479. (D) Observe that $\displaystyle\int f(x)dx = -\frac{1}{1+x} = -\frac{1}{1-(-x)} = -\sum_{n=0}^{\infty}(-1)^n x^n + c$, when $|x| < 1$. Therefore,

$$f(x) = \frac{d}{dx}\left(-\sum_{n=0}^{\infty}(-1)^n x^n + c\right) = -\sum_{n=0}^{\infty}\left((-1)^n \frac{d}{dx}(x^n)\right) + \frac{d}{dx}(c)$$

$$= (-1)^2 \sum_{n=1}^{\infty}(-1)^{n-1} nx^{n-1} = \sum_{n=0}^{\infty}(-1)^n(n+1)x^n$$

with the same radius of convergence $r = 1$.

480. (A) Observe that when $x \neq 0$

$$\sqrt[n]{|n^n x^{2n}|} = n|x|^2 \to \infty \quad \text{as} \quad n \to \infty$$

Therefore, by the root test, the series diverges for all $x \neq 0$. It converges trivially when $x = 0$.

481. (B) Observe that s is a geometric series: $s = 10 \sum_{n=1}^{\infty} \left(\frac{2}{5}\right)^{n-1}$. You are asked to find s_{10}, the sum of the first 10 terms. You use the formula for computing a partial sum of a geometric series:

$$s_{10} = \frac{10\left(1 - \left(\frac{2}{5}\right)^{10}\right)}{1 - \frac{2}{5}} \approx 16.667$$

482. (C) You use the well-known Maclaurin series of e^x:

$$e^x = \sum_{n=0}^{\infty} \frac{x^n}{n!}$$

with radius of convergence $r = \infty$. Replacing x by $2x$ in Eq. (2), multiplying by x^2, and simplifying, we obtain

$$x^2 e^{2x} = x^2 \sum_{n=0}^{\infty} \frac{(2x)^n}{n!} = x^2 \sum_{n=0}^{\infty} \frac{2^n}{n!} x^n = \sum_{n=0}^{\infty} \frac{2^n}{n!} x^{n+2}$$

The radius of convergence remains $r = \infty$.

483. (E) You are asked to find $\sum_{n=0}^{2} \frac{f^{(n)}(0)}{n!} x^n$. Simplifying first with properties of logarithms, you see $f(x) = \ln(x+1) - 2\ln(x+2)$. Then

$$f'(x) = \frac{1}{x+1} - \frac{2}{x+2} \quad \text{and} \quad f''(x) = -\frac{1}{(x+1)^2} + \frac{2}{(x+2)^2}$$

Next, you evaluate at $x = 0$, yielding:

$$f(0) = -2\ln 2, \quad f'(0) = 1 - 1 = 0, \quad \text{and} \quad f''(0) = -1 + \frac{1}{2} = -\frac{1}{2}$$

Therefore, the degree 2 Maclaurin polynomial is found to be

$$\sum_{n=0}^{2} \frac{f^{(n)}(0)}{n!} x^n = \frac{-2\ln 2}{0!} x^0 + \frac{0}{1!} x^1 + \frac{\left(-\frac{1}{2}\right)}{2!} x^2 = -2\ln 2 - \frac{1}{4} x^2$$

484. (A) The Taylor series for f centered at $x=2$ is of the form $f(x) = \displaystyle\sum_{n=0}^{\infty} \frac{f^{(n)}(2)}{n!}(x-2)^n$. Therefore, you equate coefficients and solve for $f'''(2)$:

$$\frac{f'''(2)}{3!} = -7 \Leftrightarrow f'''(2) = -42$$

485. (D) Your plan is to use the well-known Maclaurin series for $\cos x$:

$$\cos x = \sum_{n=0}^{\infty} \frac{(-1)^n x^{2n}}{(2n)!} \tag{3}$$

First, note that $\cos^2(x) = \dfrac{1+\cos(2x)}{2}$. Therefore, replacing x by $2x$ in Eq. (3), you obtain

$$\frac{1+\cos(2x)}{2} = \frac{1}{2} + \frac{1}{2}\sum_{n=0}^{\infty} \frac{(-1)^n (2x)^{2n}}{(2n)!} = \frac{1}{2} + \sum_{n=0}^{\infty} \frac{(-1)^n 2^{2n-1}}{(2n)!} x^{2n}$$

486. (B) Recall the Maclaurin series of e^x:

$$e^x = \sum_{n=0}^{\infty} \frac{x^n}{n!}$$

with radius of convergence $r = \infty$.

You replace x by $-x$ in the Maclaurin series and simplify, yielding

$$\frac{e^{-x}-1}{x} = \frac{1}{x}\left(-1 + \sum_{n=0}^{\infty} \frac{(-x)^n}{n!}\right) = \frac{1}{x}\left(-1 + \sum_{n=0}^{\infty} \frac{(-1)^n x^n}{n!}\right) = \frac{1}{x}\left(-1 + 1 + \sum_{n=1}^{\infty} \frac{(-1)^n x^n}{n!}\right)$$

$$= \frac{1}{x}\left(\sum_{n=1}^{\infty} \frac{(-1)^n x^n}{n!}\right) = \sum_{n=1}^{\infty} \frac{(-1)^n x^{n-1}}{n!}$$

Next, you integrate to obtain

$$\int_0^1 \left(\sum_{n=1}^{\infty} \frac{(-1)^n x^{n-1}}{n!}\right) dx = \sum_{n=1}^{\infty} \left(\frac{(-1)^n}{n!} \int_0^1 x^{n-1} dx\right) = \sum_{n=1}^{\infty} \left(\frac{(-1)^n}{n!}\left[\frac{x^n}{n}\right]_0^1\right)$$

$$= \sum_{n=1}^{\infty} \left(\frac{(-1)^n}{n!}\left[\frac{1}{n} - 0\right]\right) = \sum_{n=1}^{\infty} \left(\frac{(-1)^n}{n \cdot n!}\right) = \sum_{n=0}^{\infty} \frac{(-1)^{n+1}}{(n+1)\cdot(n+1)!}$$

This is an alternating series, which you will denote by s. Let $b_n = \dfrac{1}{(n+1)\cdot(n+1)!}$, and denote by s_n the nth partial sum. When estimating alternating series, you use that $|s - s_n| < b_{n+1}$. Hence,

$$|s - s_n| \le \frac{1}{(n+2)\cdot(n+2)!}$$

Observe that when $n = 5$, $|s - s_5| \le \dfrac{1}{7 \cdot 7!} \approx 0.00002$. Computing

$$s_5 = -1 + \frac{1}{2 \cdot 2!} - \frac{1}{3 \cdot 3!} + \frac{1}{4 \cdot 4!} - \frac{1}{5 \cdot 5!} + \frac{1}{6 \cdot 6!} \approx -0.7965$$

Observe that an error of 0.00002 does not affect the third place after the decimal, so your approximation -0.796 is correct to three decimal places.

487. (A) The Taylor series centered at $x = 4$ is of the form $f(x) = \displaystyle\sum_{n=0}^{\infty} \frac{f^{(n)}(4)}{n!}(x-4)^n$.

Therefore, you compute the general nth derivative $f^{(n)}(4)$. To do this, you first compute enough derivatives to find a pattern:

$$f(x) = x^{-1/2} = \frac{1}{\sqrt{x}}; \qquad\qquad f(4) = \frac{1}{2}$$

$$f'(x) = \frac{-1}{2}x^{-3/2} = -\frac{1}{2\left(\sqrt{x}\right)^3}; \qquad f'(4) = -\frac{1}{2^4}$$

$$f''(x) = \frac{1 \cdot 3}{2^2}x^{-5/2} = \frac{1 \cdot 3}{2^2\left(\sqrt{x}\right)^5}; \qquad f''(4) = \frac{1 \cdot 3}{2^7}$$

$$f'''(x) = -\frac{1 \cdot 3 \cdot 5}{2^3}x^{-7/2} = -\frac{1 \cdot 3 \cdot 5}{2^3\left(\sqrt{x}\right)^7}; \quad f'''(4) = -\frac{1 \cdot 3 \cdot 5}{2^{10}}$$

$$f^{(4)}(x) = \frac{1 \cdot 3 \cdot 5 \cdot 7}{2^4}x^{-9/2} = \frac{1 \cdot 3 \cdot 5 \cdot 7}{2^4\left(\sqrt{x}\right)^9}; \quad f^{(4)}(4) = \frac{1 \cdot 3 \cdot 5 \cdot 7}{2^{13}}$$

The pattern which emerges is evidently: $f^{(n)}(4) = (-1)^n \dfrac{1 \cdot 3 \cdot 5 \cdots (2n-1)}{2^{3n+1}}$. Hence,

$$f(x) = \sum_{n=0}^{\infty} (-1)^n \frac{1 \cdot 3 \cdot 5 \cdots (2n-1)}{2^{3n+1} n!}(x-4)^n$$

488. (B) The Taylor series centered at $x = 1$ is of the form $f(x) = \displaystyle\sum_{n=0}^{\infty} \frac{f^{(n)}(1)}{n!}(x-1)^n$.
In order to find $f^{(n)}(1)$, you compute a few derivatives:

$$f(x) = \cos(\pi x), \qquad\quad f'(x) = -\pi\sin(\pi x),$$
$$f''(x) = -\pi^2\cos(\pi x), \qquad f'''(x) = +\pi^3\sin(\pi x),$$
$$f^{(4)}(x) = +\pi^4\cos(\pi x), \qquad f^{(5)}(x) = -\pi^5\sin(\pi x).$$

Observe that when you evaluate at $x = 1$, you obtain

$$f^{(n)}(1) = \begin{cases} (-1)^{k+1}\pi^{2k} & \text{if} \quad n = 2k \\ 0 & \text{if} \quad n = 2k-1 \end{cases}$$

Therefore, the Taylor series centered at $x = 1$ for $\cos(\pi x)$ is

$$\cos(\pi x) = \sum_{n=0}^{\infty} \frac{(-1)^{n+1} \pi^{2n}}{(2n)!} (x-1)^{2n}$$

489. (E) Recall the Maclaurin series of e^x:

$$e^x = \sum_{n=0}^{\infty} \frac{x^n}{n!}$$

with radius of convergence $r = \infty$.

Replacing x by $-\dfrac{2}{3}$ in the above series, you obtain

$$e^{-2/3} = \sum_{n=0}^{\infty} \frac{\left(-\dfrac{2}{3}\right)^n}{n!} = \sum_{n=0}^{\infty} \frac{(-1)^n 2^n}{3^n n!}$$

Hence, the sum of the series is the value $e^{-2/3} \approx 0.513417$.

490. (C) First, observe that $f(x) = e^x \cdot \left(\dfrac{1}{1-x}\right)$. The Maclaurin series for $\dfrac{1}{1-x} = \sum_{n=0}^{\infty} x^n$ is the geometric sum formula, and that of e^x is also well-known:

$$e^x = \sum_{n=0}^{\infty} \frac{x^n}{n!}$$

Expanding the each series in the product yields

$$e^x \cdot \left(\frac{1}{1-x}\right) = \left(1 + x + \frac{x^2}{2} + \cdots\right)(1 + x + x^2 + \cdots)$$

Multiplying this out and collecting like terms,

$$
\begin{aligned}
&1 + x + \frac{x^2}{2} + \cdots \\
&1 + x + x^2 + \cdots \\
\hline
&1 + x + \frac{x^2}{2} + \cdots \\
&\quad\; x + x^2 + \frac{x^3}{2} + \cdots \\
&\qquad\quad x^2 + x^3 + \frac{x^4}{2} + \cdots \\
\hline
&1 + 2x + \frac{5}{2}x^2 + \cdots
\end{aligned}
$$

Therefore,

$$\frac{e^x}{1-x} = 1 + 2x + \frac{5}{2}x^2 + \cdots$$

491. (E) Recall the Maclaurin series for $\cos x$:

$$\cos x = \sum_{n=0}^{\infty} \frac{(-1)^n x^{2n}}{(2n)!}$$

Next, observe that $1° = \dfrac{\pi}{20}$. Then, you replace x by $\dfrac{\pi}{20}$ in the Maclaurin series for $\cos x$ to obtain

$$\cos\left(\frac{\pi}{20}\right) = 1 - \frac{(\pi/20)^2}{2!} + \frac{(\pi/20)^4}{4!} - \frac{(\pi/20)^6}{6!} + \cdots$$

For your approximation, you use Taylor's inequality, which says that since $f^{(n+1)}(x) \le 1$ for all x, then the difference between the sum $s(x)$ and the nth partial sum $s_n(x)$ is given by

$$\left| s\left(\frac{\pi}{20}\right) - s_n\left(\frac{\pi}{20}\right) \right| \le \frac{1}{(n+1)!} \left(\frac{\pi}{20}\right)^{n+1}$$

This error is less than or equal to 0.0000253 when $n \ge 3$. Therefore,

$$\cos\left(\frac{\pi}{20}\right) \approx 1 - \frac{(\pi/20)^2}{2!} + \frac{(\pi/20)^4}{4!} - \frac{(\pi/20)^6}{6!} \approx 0.98768$$

492. (B) Recall the Maclaurin series for $\sin x$:

$$\sin x = \sum_{n=0}^{\infty} \frac{(-1)^n}{(2n+1)!} x^{2n+1}$$

Next, observe that the given series $\dfrac{\pi}{3} - \dfrac{(\pi/3)^3}{3!} + \dfrac{(\pi/3)^5}{5!} - \dfrac{(\pi/3)^7}{7!} + \cdots$ may be rewritten as

$$\sum_{n=0}^{\infty} \frac{(-1)^n}{(2n+1)!} \left(\frac{\pi}{3}\right)^{2n+1}$$

That is, the given infinite series is the result of replacing x by $\dfrac{\pi}{3}$ in the Maclaurin series for $\sin x$. Hence, the sum of the series is $\sin \dfrac{\pi}{3} = \dfrac{\sqrt{3}}{2}$.

493. (E) Recall the Maclaurin series for $\cos x$:

$$\cos x = \sum_{n=0}^{\infty} \frac{(-1)^n x^{2n}}{(2n)!}$$

Replacing x by x^3 in the formula yields

$$\cos(x^3) = \sum_{n=0}^{\infty} \frac{(-1)^n (x^3)^{2n}}{(2n)!} = \sum_{n=0}^{\infty} \frac{(-1)^n x^{6n}}{(2n)!} = 1 + \sum_{n=1}^{\infty} \frac{(-1)^n x^{6n}}{(2n)!}$$

Therefore, $\cos(x^3) - 1 = \sum_{n=1}^{\infty} \frac{(-1)^n x^{6n}}{(2n)!}$, and when you divide by x, you obtain

$$\frac{\cos(x^3) - 1}{x} = \frac{1}{x}(\cos(x^3) - 1) = \frac{1}{x} \sum_{n=1}^{\infty} \frac{(-1)^n x^{6n}}{(2n)!} = \sum_{n=1}^{\infty} \frac{(-1)^n x^{6n-1}}{(2n)!}$$

Next, you integrate term-by-term, to conclude:

$$\int \left(\frac{\cos(x^3) - 1}{x} \right) dx = \int \left(\sum_{n=1}^{\infty} \frac{(-1)^n x^{6n-1}}{(2n)!} \right) dx = \sum_{n=1}^{\infty} \left(\frac{(-1)^n}{(2n)!} \int x^{6n-1} \, dx \right)$$

$$= \sum_{n=1}^{\infty} \frac{(-1)^n}{(2n)!} \left(\frac{x^{6n}}{6n} \right) + c = \sum_{n=1}^{\infty} \frac{(-1)^n}{6n(2n)!} x^{6n} + c$$

for some constant c, and you have evaluated and indefinite integral as an infinite series.

494. (C) One solution is to consider $f(x) = \ln(1 - x)$, and notice that

$$\ln(1 - x) = -\int \frac{1}{1-x} dx = -\sum_{n=0}^{\infty} \left(\int x^n dx \right) = -\sum_{n=1}^{\infty} \frac{x^n}{n}$$

(The constant of integration is 0, upon evaluation at $x = 0$.)

Therefore, $T_4(x) = -x - \frac{1}{2}x^2 - \frac{1}{3}x^3 + \frac{1}{4}x^4$, and

$$\ln(2) = \ln(1 - (-1)) = f(-1) \approx T_2(-1) = 1 - \frac{1}{2} + \frac{1}{3} - \frac{1}{4} = \frac{7}{12}$$

495. (D) The series in I is geometric, hence you use Equation (1), replacing x by $\frac{1}{3}$ to obtain

$$2 \sum_{n=0}^{\infty} \left(\frac{1}{3} \right)^n = 2 \frac{1}{\left(1 - \frac{1}{3} \right)} = 2 \cdot \left(\frac{3}{2} \right) = 3$$

Note that in II, you have simply replaced x by $\ln 3$ in the Maclaurin series for e^x in Eq. (2). Thus,

$$1 + \ln 3 + \frac{(\ln 3)^2}{2} + \frac{(\ln 3)^3}{6} + \frac{(\ln 3)^4}{24} + \cdots = \sum_{n=0}^{\infty} \frac{(\ln 3)^n}{n!} = e^{\ln 3} = 3$$

The series in III diverges, since $\frac{3n+1}{n-8} \not\to 0$ as $n \to \infty$.

496. (A) You use the ratio test to test convergence. Let $a_n = \dfrac{n^n}{(3n)!}$. Then

$$\left|\frac{a_{n+1}}{a_n}\right| = \frac{(n+1)^{n+1}(3n)!}{(3n+3)!n^n} = (n+1)\left(\frac{n+1}{n}\right)^n \frac{1}{(3n+3)(3n+2)(3n+1)}$$

$$= \frac{n+1}{3n+3}\left(\frac{n+1}{n}\right)^n \frac{1}{(3n+2)(3n+1)} = \frac{1}{3}\left(1+\frac{1}{n}\right)^n \frac{1}{(3n+2)(3n+1)}$$

Recall the well-known limit:

$$\lim_{n\to\infty}\left(1+\frac{1}{n}\right)^n = e$$

Then, you let $n \to \infty$ to find that

$$\lim_{n\to\infty}\left|\frac{a_{n+1}}{a_n}\right| = \lim_{n\to\infty}\frac{1}{3}\left(1+\frac{1}{n}\right)^n \frac{1}{(3n+2)(3n+1)} = \frac{1}{3}\cdot e\cdot 0 = 0 < 1$$

Hence, by the ratio test, the series converges.

(B) As the series converges, the terms must limit to zero, that is, $\lim_{n\to\infty}\dfrac{n^n}{(3n)!} = 0$.

497. (A) You are looking to write $f(x) \approx T_3(x) = \displaystyle\sum_{n=0}^{3}\frac{f^{(n)}(1)}{n!}(x-1)^n$, so you must compute the derivatives $f^{(n)}(1)$ for $n = 0, 1, 2, 3$:

$$f(x) = x^{-2} = \frac{1}{x^2} \qquad\qquad f(1) = 1$$

$$f'(x) = -2x^{-3} = \frac{-2}{x^3} \qquad\qquad f'(1) = -2$$

$$f''(x) = 6x^{-4} = \frac{6}{x^4} \qquad\qquad f''(1) = 6$$

$$f'''(x) = -24x^{-5} = \frac{-24}{x^5} \qquad\qquad f'''(1) = -24$$

Therefore,

$$T_3(x) = \sum_{n=0}^{3}\frac{f^{(n)}(1)}{n!}(x-1)^n = 1 - \frac{2}{1!}(x-1) + \frac{6}{2!}(x-1)^2 - \frac{24}{3!}(x-1)^3$$

$$= 1 - 2(x-1) + 3(x-1)^2 - 4(x-1)^3$$

(B) Write $f(x) = T_3(x) + R_3(x)$. Taylor's inequality states: if $f^4(x) \le M$ for $x \in [0.8, 1.2]$, then

$$|R_3(x)| \le \frac{M}{4!}|x-1|^4 \quad \text{for} \quad x \in [0.8, 1.2]$$

You begin by computing $f^{(4)}(x) = 120x^{-6} = \dfrac{120}{x^6}$. Observe that $f^{(4)}(x)$ achieves its maximum on $[0.8, 1.2]$ at $x = 0.8 = \dfrac{4}{5}$. That is,

$$f^{(4)}(x) = \frac{120}{x^6} \le \frac{120}{(4/5)} = \frac{120}{1} \cdot \frac{5}{4} = 30 \cdot 5 = 150$$

Furthermore, note that $|x-1| \le 0.2 = \dfrac{1}{5}$ on $[0.8, 1.2]$.

Now, you are able to find the maximum possible error $|f(x) - T_3(x)| = |R_3(x)|$ on the interval $[0.8, 1.2]$:

$$|R_3(x)| \le \frac{150}{4!}|x-1|^4 \le \frac{150}{4!}\left(\frac{1}{5}\right)^4 = \frac{\cancel{25} \cdot 6}{24}\left(\frac{1}{25}\right)^{\cancel{2}}$$

$$= \frac{1}{4 \cdot 25} = \frac{1}{100} \quad \text{for} \quad x \in [0.8, 1.2]$$

498. (A) The answer follows from the observation that

$$a_4 = (a_1 + a_2 + a_3 + a_4) - (a_1 + a_2 + a_3) = s_4 - s_3$$

Since $s_4 = 4\tan\left(\dfrac{\pi}{4}\right) = 4 \cdot 1 = 4$ and $s_3 = 3\tan\left(\dfrac{\pi}{3}\right) = 3\dfrac{1}{\sqrt{3}} = \sqrt{3}$, you find that

$$a_4 = 4 - \sqrt{3}.$$

(B) Next, you use definition of a convergent series, in that the sum of the infinite series is the limit of its partial sums. That is:

$$\sum_{n=0}^{\infty} a_n = \lim_{N \to \infty} \sum_{n=0}^{N} a_n = \lim_{N \to \infty} s_N = \lim_{N \to \infty} n\tan\left(\frac{\pi}{n}\right)$$

Note that this limit has the indeterminate form $\infty \cdot 0$. Thus, in order to compute this limit, you rewrite it as $\lim_{N \to \infty} \dfrac{\tan(\pi/n)}{(1/n)}$. This has the indeterminate form $\dfrac{0}{0}$, hence satisfies the hypotheses for l'Hospital's rule. You may now compute the limit:

$$\lim_{N \to \infty} \frac{\tan(\pi/N)}{(1/N)} = \lim_{N \to \infty} \frac{\sec^2(\pi/N) \cdot \pi \cancel{(-N^{-2})}}{\cancel{(-N^{-2})}} = \pi \sec^2(0) = \pi$$

499. (A) Let $a_n = c_n(-3)^n$, and consider the following limit:

$$\lim_{n \to \infty} \left| \frac{a_{n+1}}{a_n} \right| = \lim_{n \to \infty} \left| \frac{c_{n+1}(-3)^{n+1}}{c_n(-3)^n} \right| = \lim_{n \to \infty} \left| \frac{c_{n+1}}{c_n} \right| \cdot |-3|$$

You are told that $\sum_{n=0}^{\infty} c_n(-3)^n$ converges, hence, by the ratio test

$$\lim_{n \to \infty} \left| \frac{c_{n+1}}{c_n} \right| \cdot |-3| \leq 1$$

Rearranging the above inequality, you obtain

$$\lim_{n \to \infty} \left| \frac{c_{n+1}}{c_n} \right| \leq \frac{1}{3}$$

And, since $\frac{1}{3} < 1$, it follows by the ratio test that $\sum_{n=0}^{\infty} c_n$ converges.

(B) Before considering the series $\sum_{n=0}^{\infty} c_n 6^n$, we show that $\lim_{n \to \infty} \left| \frac{c_{n+1}}{c_n} \right| \geq \frac{1}{5}$.

Let $b_n = c_n 5^n$, and consider the following limit:

$$\lim_{n \to \infty} \left| \frac{b_{n+1}}{b_n} \right| = \lim_{n \to \infty} \left| \frac{c_{n+1} 5^{n+1}}{c_n 5^n} \right| = \lim_{n \to \infty} \left| \frac{c_{n+1}}{c_n} \right| \cdot 5$$

You are told that $\sum_{n=0}^{\infty} c_n 5^n$ diverges, hence, by the ratio test

$$\lim_{n \to \infty} \left| \frac{c_{n+1}}{c_n} \right| \cdot 5 \geq 1$$

Rearranging the above inequality, you obtain

$$\lim_{n \to \infty} \left| \frac{c_{n+1}}{c_n} \right| \geq \frac{1}{5}$$

Now, let $d_n = c_n 6^n$ and consider the following limit:

$$\lim_{n \to \infty} \left| \frac{d_{n+1}}{d_n} \right| = \lim_{n \to \infty} \left| \frac{c_{n+1} 6^{n+1}}{c_n 6^n} \right| = \lim_{n \to \infty} \left| \frac{c_{n+1}}{c_n} \right| \cdot 6$$

Note that by Eq. (1)

$$\lim_{n \to \infty} \left| \frac{c_{n+1}}{c_n} \right| \cdot 6 \geq \frac{6}{5} > 1$$

By the ratio test, you conclude that $\sum_{n=0}^{\infty} c_n 6^n$ diverges.

(C) $\displaystyle\sum_{n=0}^{\infty} c_n(-2)^n$

Let $e_n = c_n(-2)^n$, and note that by Eq. (1):

$$\lim_{n\to\infty}\left|\frac{e_{n+1}}{e_n}\right| = \lim_{n\to\infty}\left|\frac{c_{n+1}(-2)^{n+1}}{c_n(-2)^n}\right| = \lim_{n\to\infty}\left|\frac{c_{n+1}}{c_n}\right|\cdot 2 \le \frac{2}{3} < 1$$

And, since $\dfrac{2}{3} < 1$, it follows by the ratio test that $\displaystyle\sum_{n=0}^{\infty} c_n(-2)^n$ converges.

(D) Now, let $f_n = c_n(-7)^n$ and consider the following limit:

$$\lim_{n\to\infty}\left|\frac{f_{n+1}}{f_n}\right| = \lim_{n\to\infty}\left|\frac{c_{n+1}(-7)^{n+1}}{c_n(-7)^n}\right| = \lim_{n\to\infty}\left|\frac{c_{n+1}}{c_n}\right|\cdot 7$$

Note that by Eq. (2)

$$\lim_{n\to\infty}\left|\frac{c_{n+1}}{c_n}\right|\cdot 7 \ge \frac{7}{5} > 1$$

By the ratio test, you conclude that $\displaystyle\sum_{n=0}^{\infty} (-1)^n c_n 7^n$ diverges.

500. (A) You use the Maclaurin series of $\sin x$:

$$\frac{\sin x}{x} = \frac{1}{x}\sum_{n=0}^{\infty}(-1)^n\frac{x^{2n+1}}{(2n+1)!} = \sum_{n=0}^{\infty}(-1)^n\frac{x^{2n}}{(2n+1)!}$$

(B) You integrate term-by-term in Eq. (6):

$$\int\frac{\sin x}{x}\,dx = \int\left(\sum_{n=0}^{\infty}(-1)^n\frac{x^{2n}}{(2n+1)!}\right)dx = \sum_{n=0}^{\infty}\left(\frac{(-1)^n}{(2n+1)!}\int x^{2n}\,dx\right)$$

$$= \left(\sum_{n=0}^{\infty}\frac{(-1)^n x^{2n+1}}{(2n+1)!(2n+1)}\right)+c$$

(C) Observe that by part (B):

$$\int_0^1\left(\frac{\sin x}{x}\right)dx = \left[\sum_{n=0}^{\infty}\frac{(-1)^n x^{2n+1}}{(2n+1)!(2n+1)}\right]_0^1 = \sum_{n=0}^{\infty}\frac{(-1)^n}{(2n+1)!(2n+1)}$$

This is an alternating series, hence the difference $|s - s_N|$ between the infinite series above and its Nth partial sum is *at most* the absolute value of the $(N+1)$th term of the series. That is,

$$\left| \sum_{n=0}^{\infty} \frac{(-1)^n}{(2n+1)!(2n+1)} - \sum_{n=0}^{N} \frac{(-1)^n}{(2n+1)!(2n+1)} \right| \leq \frac{1}{(2N+3)!(2N+3)}$$

Note that when $N \geq 2$,

$$\frac{1}{(2N+3)!(2N+3)} \leq 0.0000283447$$

And, since

$$s_2 = 0.946\overline{1}$$

a difference of 0.0000283447 will not affect the third place after the decimal. Therefore,

$$\int_0^1 \frac{\sin x}{x}\, dx = s \approx s_2 = 0.946\overline{1}$$